ε Epsilon

Fractions

Instruction Manual

By Steven P. Demme

Math·U·See.

1-888-854-MATH (6284) - www.MathUSee.com
Sales@MathUSee.com

Epsilon Instruction Manual: Fractions
©2012 Math-U-See, Inc.
Published and distributed by Demme Learning

www.MathUSee.com

1-888-854-6284 or +1 717-283-1448 | www.demmelearning.com
Lancaster, Pennsylvania USA

ISBN 978-1-60826-083-6

Printed in the United States of America

4 5 6 7 8 9 10   17 16   1014

Building Understanding in Teachers and Students to Nurture a Lifelong Love of Learning

At Math-U-See, our goal is to build understanding for all students.

We believe that education should be relevant, skills-based, and built on previous learning. Because students have a variety of learning styles, we believe education should be multi-sensory. While some memorization is necessary to learn math facts and formulas, students also must be able to apply this knowledge in real-life situations.

Math-U-See is proud to partner with teachers and parents as we use these principles of education to **build lifelong learners.**

CURRICULUM SEQUENCE

Calculus

PreCalculus
with Trigonometry

Algebra 2

Geometry

Algebra 1

Pre-Algebra

Zeta
Decimals and Percents

Epsilon
Fractions

Delta
Division

Gamma
Multiplication

Beta
Multiple-Digit Addition and Subtraction

Alpha
Single-Digit Addition and Subtraction

Primer
Introducing Math

Math-U-See is a complete, comprehensive K-12 math curriculum that uses manipulatives to illustrate and teach math concepts. We strive toward "Building Understanding" by using a multi-sensory, mastery-based approach suitable for all levels and learning styles. While each book concentrates on a specific theme, other math topics are introduced where appropriate. Subsequent books continuously review and integrate topics and concepts presented in previous levels.

Where to Start
Because Math-U-See is mastery-based, students may start at any level. We use the Greek alphabet to show the sequence of concepts taught rather than the grade level. Go to MathUSee.com for more placement help.

Each level builds on previously-learned skills to prepare a solid foundation so the student is then ready to apply these concepts to Algebra and other upper-level courses.

Major concepts and skills:

- Recognizing and generating equivalent fractions
- Understanding addition, subtraction, multiplication, and division of fractions and mixed numbers
- Fluently adding, subtracting, multiplying, and dividing fractions and mixed numbers

Additional concepts and skills:

- Using multiple strategies to recognize common factors
- Understanding grouping symbols and their effect on order of operations
- Interpreting and solving word problems
- Comparing and converting decimal fractions
- Finding the area and circumference of circles
- Classifying quadrilaterals
- Representing fractions and fractional measurements on line plots and number lines
- Using coordinates to represent ordered relationships

Find more information and products at MathUSee.com

Contents

HOW TO USE

Five Minutes for Success

Welcome to *Epsilon*. I believe you will have a positive experience with the unique Math-U-See approach to teaching math. These first few pages explain the essence of this methodology, which has worked for thousands of students and teachers. I hope you will take five minutes and read through these steps carefully.

I am assuming your student has a thorough grasp of the four basic operations: addition, subtraction, multiplication, and division.

If you are using the program properly and still need additional help, you may visit us online at MathUSee.com or call us at 888-854-6284. –**Steve Demme**

The Goal of Math-U-See

The underlying assumption or premise of Math-U-See is that the reason we study math is to apply math in everyday situations. Our goal is to help produce confident problem solvers who enjoy the study of math. These are students who learn their math facts, rules, and formulas and are able to use this knowledge to solve word problems and real-life applications. Therefore, the study of math is much more than simply committing to memory a list of facts. It includes memorization, but it also encompasses learning the underlying concepts of math that are critical to successful problem solving.

Support and Resources

Math-U-See has a number of resources to help you in the educational process.

Many of our customer service representatives have been with us for over 10 years. They are able to answer your questions, help you place your student in the appropriate level, and provide knowledgeable support throughout the school year.

Visit MathUSee.com to use our many online resources, find out when we will be in your neighborhood, and connect with us on social media.

More than Memorization

Many people confuse memorization with understanding. Once while I was teaching seven junior high students, I asked how many pieces they would each receive if there were fourteen pieces. The students' response was, "What do we do: add, subtract, multiply, or divide?" Knowing how to divide is important, but understanding when to divide is equally important.

The Suggested 4-Step Math-U-See Approach

In order to train students to be confident problem solvers, here are the four steps that I suggest you use to get the most from the Math-U-See curriculum.

Step 1. Prepare for the lesson
Step 2. Present and explore the new concept together
Step 3. Practice for mastery
Step 4. Progress after mastery

Step 1. Prepare for the lesson

Watch the video lesson to learn the new concept and see how to demonstrate this concept with the manipulatives when applicable. Study the written explanations and examples in the instruction manual.

Step 2. Present and explore the new concept together

Present the new concept to your student. Have the student watch the video lesson with you, if you think it would be helpful. The following should happen interactively.

a. **Build:** Use the manipulatives to demonstrate and model problems from the instruction manual. If you need more examples, use the appropriate lesson practice pages.

b. **Write:** Write down the step-by-step solutions as you work through the problems together, using manipulatives.

c. **Say:** Talk through the why of the math concept as you build and write.

Give as many opportunities for the student to "Build, Write, Say" as necessary until the student fully understands the new concept and can demonstrate it to you confidently. One of the joys of teaching is hearing a student say, *"Now I get it!"* or *"Now I see it!"*

Step 3. Practice for mastery

Using the lesson practice problems from the student workbook, have students practice the new concept until they understand it. It is one thing for students to watch someone else do a problem; it is quite another to do the same problem themselves. Together complete as many of the lesson practice pages as necessary (not all pages may be needed) until the student understands the new concept, demonstrating confident mastery of the skill. Remember, to demonstrate mastery, your student should be able to teach the concept back to you using the Build, Write, Say method. Give special attention to the word problems, which are designed to apply the concept being taught in the lesson. If your student needs more assistance, go to MathUSee.com to find review tools and other resources.

Step 4. Progress after mastery

Once mastery of the new concept is demonstrated, advance to the systematic review pages for that lesson. These worksheets review the new material as well as provide practice of the math concepts previously studied. If the student struggles, reteach these concepts to maintain mastery. If students quickly demonstrate mastery, they may not need to complete all of the systematic review pages.

In the 2012 student workbook, the last systematic review page for each lesson is followed by a page called "Application and Enrichment." These pages provide a way for students to review and use their math skills in a variety of different formats. Some of the Application and Enrichment pages introduce terms and ideas that a student may encounter on standardized tests. Mastery of these concepts is *not* necessary in order to move to the next level of Math-U-See. You may decide how useful these activity pages are for your particular student.

Now you are ready for the lesson tests. These were designed to be an assessment tool to help determine mastery, but they may also be used as extra worksheets. Your student will be ready for the next lesson only after demonstrating mastery of the new concept and maintaining mastery of concepts found in the systematic review worksheets.

Tell me, I forget. Show me, I understand. Let me do it, I remember.
–Ancient Proverb

To this Math-U-See adds, *"Let me teach it, and I will have achieved mastery!"*

Length of a Lesson

How long should a lesson take? This will vary from student to student and from topic to topic. You may spend a day on a new topic, or you may spend several days. There are so many factors that influence this process that it is impossible to predict the length of time from one lesson to another. I have spent three days on a lesson, and I have also invested three weeks in a lesson. This experience occurred in the same book with the same student. If you move from lesson to lesson too quickly without the student demonstrating mastery, the student will become overwhelmed and discouraged as he or she exposed to more new material without having learned previous topics. If you move too slowly, the student may become bored and lose interest in math. I believe that as you regularly spend time working along with the student, you will sense the right time to take the lesson test and progress through the book.

By following the four steps outlined above, you will have a much greater opportunity to succeed. Math must be taught sequentially, as it builds line upon line and precept upon precept on previously-learned material. I hope you will try this methodology and move at the student's pace. As you do, I think you will be helping to create a confident problem solver who enjoys the study of math.

Fraction of a Number
Word Problem Tips

Three steps to understanding fractions are introduced in this lesson. These steps will be the basis for much of our study of fractions. If you have the Math-U-See blocks, you may use the unit blocks for this lesson. If you don't have the blocks, you may use any small objects, such as raisins or buttons.

The first step is often left out, but it is important because it determines our starting point. A fraction is a fraction of something. I ask students, "Which is greater: one half or one fourth?" They usually reply, "One half," when what they should say is, "One half of what and one fourth of what?" If I then ask, "One half of this room, or one fourth of the state?", the answer will be different. Therefore, the first step is determining the beginning area or number. One half of four is different than one half of eight.

The second step is determining the ***denominator***. This is the number of equal parts into which we divide the starting number. Notice that both "divide" and "denominator" begin with *d*. The symbolism also shows this step, since the line separating the the two parts of the fraction means "divided by." We are dividing the amount from Step 1 by the the denominator.

In the third step, we count how many equal parts are indicated by the ***numerator***, or top number. I call this the "numBerator" at first to make the connection with counting. Later we take out the B and have the word numerator. A ***unit fraction*** is a fraction with a numerator of 1 that indicates one equal part.

Here is a summary of the steps.

1. Identify the the starting number.

2. Identify the denominator, which tells how many equal parts are in the starting number.

3. Identify the numBerator, or numerator, which tells us how many of these equal parts we will count.

Example 1
Find 2/3 of 6.

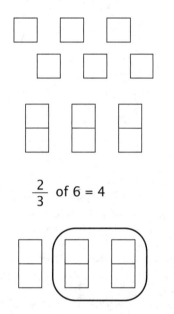

Step 1
Select 6 green unit blocks.

Step 2
Divide 6 into 3 equal parts. Each part is 1/3 (one third) of the starting number.

$\frac{2}{3}$ of 6 = 4

Step 3
Count 2 of those parts. Together, those parts are 2/3 (two thirds) of the starting number.

A *fraction* of a number is a combination of a division problem and a multiplication problem. First we divide by the denominator, and then we multiply by the numerator. In Example 1, we divided 6 by 3 to find the equal parts ($6 \div 3 = 2$). Then we multiplied the result by 2 to find the amount that is 2/3 of 6 ($2 \times 2 = 4$).

Point out to students that taking a fraction of a whole number always results in a number that is less (assuming that the fraction is less than one). This concept may seem obvious, but understanding it will help in lesson 9, where we relate the "fraction of a number" language to multiplication by a fraction.

Example 2
Find 3/4 of 12.

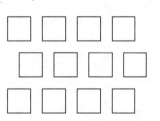

Step 1
Select 12 green unit blocks.

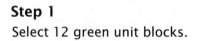

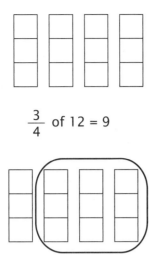

$\frac{3}{4}$ of 12 = 9

Step 2

Divide 12 into 4 equal parts. Each part is 1/4 (one fourth) of the starting number.

Step 3

Count 3 of those parts. Together, those parts are 3/4 (three fourths) of the starting number.

Word Problem Tips

Parents often find it challenging to teach children how to solve word problems. Here are some suggestions for helping your student learn this important skill.

The first step is to realize that word problems require both reading and math comprehension. Don't expect a child to be able to solve a word problem if he does not thoroughly understand the math concepts involved. On the other hand, a student may have a math skill level that is stronger than his or her reading comprehension skills. Here are a number of strategies to improve comprehension skills in the context of story problems. You may decide which ones work best for you and your child.

Strategies for Word Problems

1. Ignore numbers at first and read the story. It may help some students to read the question aloud. Every word problem tells a story. Before deciding what math operation is required, let the student retell the story in his own words. Who is involved? Are they receiving gifts, losing something, or dividing a treat?

2. Relate the story to real life, perhaps by using names of family members. This may make the problem more interesting and relevant.

3. Build, draw, or act out the story. Use the blocks or actual objects when practical. Especially in the lower levels, you may require the student

to use the blocks for word problems even when the facts have been learned. Don't be afraid to use a little drama as well. The purpose is to make it as real and meaningful as possible.

4. Look for the common language used in a particular kind of problem. Pay close attention to the word problems on the lesson practice pages, as they model different kinds of language that may be used for the new concept just studied. For example, "altogether" usually indicates addition. These "key words" can be useful clues but should not be a substitute for understanding.

5. Look for practical applications that use the concept and ask questions in that context.

6. Have the student invent word problems to illustrate his number problems from the lesson.

Cautions

1. Unneeded information may be included in the problem. For example, we may be told that Suzie is eight years old, but the eight is irrelevant when adding up the number of gifts she received.

2. Some problems may require more than one step to solve. Model these questions carefully.

3. There may be more than one way to solve some problems. Experience will help the student choose the easier or preferred method.

4. Estimation is a valuable tool for checking an answer. If an answer is unreasonable, it is possible that the wrong method was used to solve the problem.

Note: The "Application and Enrichment" pages in the *Epsilon Student Workbook* include many activities designed to help students understand and solve word problems involving fractions.

Fraction of One

In lesson 1, we learned three steps for finding the fraction of a number. Now we will apply the same steps to finding a fraction of one.

I've been using a green unit block to represent one. To transition from the blocks to the overlays, I select the green 5×5 inch square overlay to represent one. Then I hold up a white square and place it over the green one, showing that they are the same size and thus have the same value—one. I prefer the white square because it is less confusing when used with the colored inserts.

Holding the white square to represent one (the first step), take the clear overlay with four vertical lines and place it on top of the square to divide it into five equal parts (the denominator). We've done two steps so far. We've taken one and divided it into five equal parts. Now, how many of those equal parts are we going to count (the numerator)? In Example 1, we want two parts, so we use the light blue 2/5 piece (which is the same as two 1/5 pieces) to represent the numerator.

Example 1
Find 2/5 of 1.

Step 1
Begin with 1.

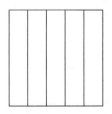

Step 2
Divide 1 into 5 equal parts. Each part is 1/5 of the whole.

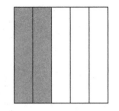

Step 3
Count 2 of those parts.

Example 2
Find 3/4 of 1.

Step 1
Begin with 1.

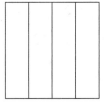

Step 2
Divide 1 into 4 equal parts. Each part is 1/4 of the whole.

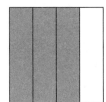

Step 3
Count 3 of those parts.

These are the same steps used to find a fraction of a number. Stress that the horizontal line separating the numerator and the denominator means "divided by." Before you begin the worksheets, read the following instructions carefully.

A. Build several different fractions with the overlays and ask the student to write what you've built. Then have him or her read the correct answer out loud. When the student is comfortable with this, reverse the roles and ask the student to build several for you to write out and say.

B. When Step A is understood, write a fraction and have the student build it and say it. Do this again with the student as the instructor.

C. Now say a fraction out loud and have the student build and write it. Then have the student say one for you to build and write.

Use this process with all new material whenever possible, even if the directions in each section do not specifically tell you to do so. Build—Write—Say and then reverse roles to model good problem solving skills for the student.

Add and Subtract Fractions
Common Denominator; Mental Math

To add or subtract fractions, we must have the same number of equal parts. With place value, we can add only tens to tens, units to units, or hundreds to hundreds. In fractions, the denominator indicates what unit or kind or value is being counted by the numerator. It is easier to add two fractions that have the same denominator because the denominator is our unit of comparison. If two fractions have the same denominator, it is called a ***common denominator***. I say "same denominator" for some time before I use the words "common denominator." This may make the meaning clearer for the student.

When we add and subtract fractions, we are joining or separating parts of a whole. Illustrate this by adding 1/4 to 2/4. Make 1/4 with the overlays. Now take the other clear fourths overlay and place it on top of the yellow piece that represents 2/4, without a white background. Now place the 2/4 on top of the 1/4. You can see that the answer is 3/4, as everything lines up. Do several of these until the student understands that, when you have the same denominator, you add the numerators. The same holds true for subtraction.

Example 1

Solve: 2/4 + 1/4

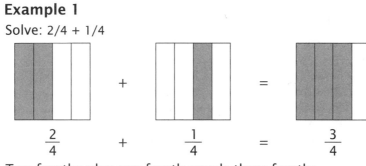

$$\frac{2}{4} \quad + \quad \frac{1}{4} \quad = \quad \frac{3}{4}$$

Two fourths plus one fourth equals three fourths.

Example 2
Solve: 2/6 + 3/6

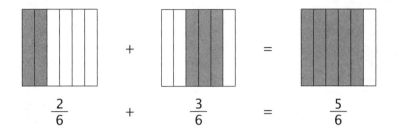

$$\frac{2}{6} + \frac{3}{6} = \frac{5}{6}$$

Two sixths plus three sixths equals five sixths.

Example 3
Solve: 5/6 – 1/6

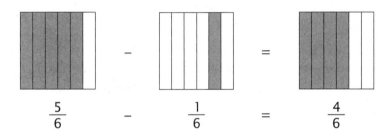

$$\frac{5}{6} - \frac{1}{6} = \frac{4}{6}$$

Five sixths minus one sixth equals four sixths.

Decomposition with Fractions

Once the student has mastered adding fractions, try giving a fraction with a numerator greater than one and challenging the student to find all the ways it can be broken down into an addition problem. For example, this problem could be rewritten as 3/6 is 1/6 + 1/6 + 1/6 or as 1/6 + 2/6. This is called *decomposing* the problem. Use the fraction overlays and write the results each time.

Mental Math

Mental math problems may be used to keep the facts fresh in the memory and to develop mental math skills. At first you will need to go slowly enough for the student to verbalize the intermediate steps. As skills increase, the student should be able to give just the answer. The purpose is to stretch but not discourage.

Example 4
Five times three, plus six, divided by seven, equals?
The student thinks: $5 \times 3 = 15$, $15 + 6 = 21$, and $21 \div 7 = 3$

Notice that the example includes an addition problem that is not a basic addition fact. These questions include addition, subtraction, multiplication, and division. Do a few at a time, remembering to go slowly at first.

1. Three plus five, times six, minus three, equals ? (45)

2. Eight times three, divided by six, plus nine, equals ? (13)

3. Nineteen minus nine, plus 10, divided by five, equals ? (4)

4. Twenty-seven divided by three, times five, plus two, equals ? (47)

5. Five plus four, times eight, plus six, equals ? (78)

6. Fifty-four divided by six, divided by three, times seven, equals ? (21)

7. Thirty-two minus two, times three, divided by 10, equals ? (9)

8. Seven minus five, times eight, divided by four, equals ? (4)

9. Seven times seven, plus one, divided by five, equals ? (10)

10. Nine plus two, times four, minus five, equals ? (39)

There are more mental math problems in lessons 9, 15, and 21. There are also mental math problems in the student workbook, starting with lesson 21D.

Equivalent Fractions

Several important skills, including adding, subtracting, dividing, and simplifying fractions, hinge on understanding equivalent fractions. This is the watershed of fractions because you must thoroughly understand this concept to be successful.

Before you begin this lesson, look through the fraction kit and notice that the colors are synchronized with the manipulative blocks. For example, the block for three is pink, and the inserts for thirds in the fraction kit are also pink.

On paper, it is difficult to see how 1/2 can be the same as 2/4 and 3/6. As you study Example 1, build it with the overlays. Begin by making 1/2. Use the white square for your background, place the clear halves overlay on it and then slide the small orange piece under the overlay. We began with one (white piece), divided it into two pieces (clear overlay), and counted one of them (orange piece). After you have made 1/2, place the other clear halves overlay horizontally on top of it to show 2/4. Then remove the second halves overlay and place the clear thirds overlay horizontally on top to show 3/6. Finally, try the fourths and fifths overlays.

Example 1

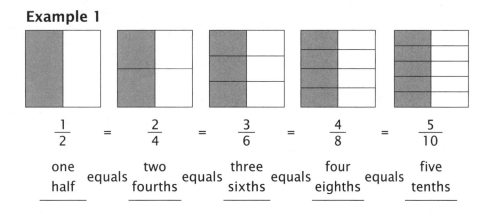

$$\frac{1}{2} = \frac{2}{4} = \frac{3}{6} = \frac{4}{8} = \frac{5}{10}$$

one half equals two fourths equals three sixths equals four eighths equals five tenths

At this point, I often tell a story. Suppose you have one half of a pizza left over from dinner last night that you were planning to eat for lunch today. Before you can eat it, the doorbell rings, and two friends drop by. You invite them in, cut the pizza into three equal parts, and give everyone one piece. After you cut the pizza and before you give it to your friends, stop and think. Do you still have the same amount of pizza? Yes. Are there more pieces? Yes. What fraction did you have originally? One half. Now what do you have (with the overlay added)? Three sixths. Do you still have the same amount? Yes. (Take off the second overlay if needed to show that it is still one half.) Do you have more pieces (or equal parts)? Yes. Keep trying different overlays and combinations until the student understands this critical concept, which I summarize by saying, "Same amount, more pieces."

Example 2

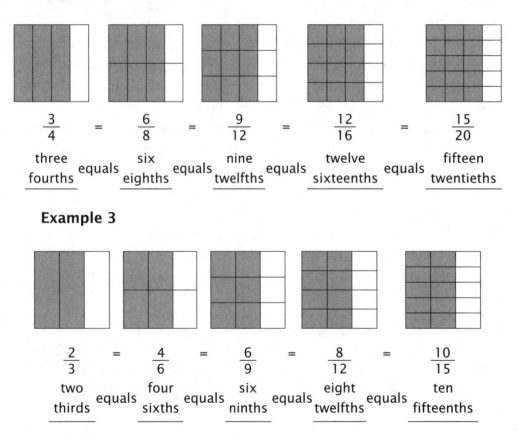

$$\frac{3}{4} = \frac{6}{8} = \frac{9}{12} = \frac{12}{16} = \frac{15}{20}$$

three fourths — equals — six eighths — equals — nine twelfths — equals — twelve sixteenths — equals — fifteen twentieths

Example 3

$$\frac{2}{3} = \frac{4}{6} = \frac{6}{9} = \frac{8}{12} = \frac{10}{15}$$

two thirds — equals — four sixths — equals — six ninths — equals — eight twelfths — equals — ten fifteenths

Addition and Subtraction
Unequal Denominators

This lesson teaches two methods for adding fractions with unequal denominators. Have the student begin by using the long method described first. When that is learned, the student should try the shortcut at the end of the lesson. Lesson practice pages 5A and 5B in the student book should be done the long way, while 5C asks the student to use the shortcut.

Because of place value, we add tens to tens, units to units, and hundreds to hundreds. When adding or subtracting fractions, we make the denominators the same so that they are the same value or unit of measure. If two fractions have the same denominator, we say they have a common denominator. We used this concept when adding fractions like 1/5 and 3/5. When the fractions had the same denominator, we added the numerators. The same principle holds true for subtraction. You may say "same denominator" instead of "common denominator" for now if it makes more sense to the student.

Try adding 2/5 and 1/3 with the overlays just as you did with 1/5 plus 3/5. When you place one fraction on top of the other, you can see that there is no clear answer. It is easier to combine them if they have the same denominator or are the same kind. In order to add or subtract two fractions, we should first find the same, or common, denominator.

Since the concept of equivalent fractions has been mastered, this kind of problem is easier because we know how to change the number of pieces while still keeping the same amount. The procedure is to keep changing the 2/5 and the 1/3 until they have the same denominator. Remember the rule: if the denominators are the same, we can combine the numerators.

We start by making the two fractions — in this case, 1/3 and 2/5. Then we begin by placing a 1/2 overlay on top of each fraction. This doesn't give us a common denominator, so we try the 1/3 overlays. We keep trying overlays until we find a common denominator. Then we can add the numerators. Study the two examples.

Example 1

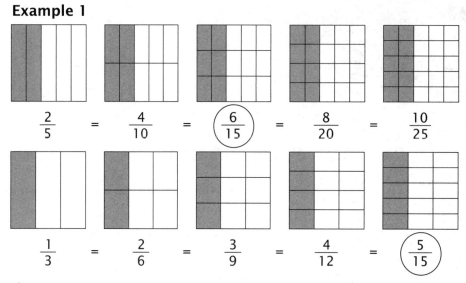

$$\frac{2}{5} = \frac{4}{10} = \boxed{\frac{6}{15}} = \frac{8}{20} = \frac{10}{25}$$

$$\frac{1}{3} = \frac{2}{6} = \frac{3}{9} = \frac{4}{12} = \boxed{\frac{5}{15}}$$

6/15 + 5/15 = 11/15

If we turn one of the fractions 90 degrees, we can see that each small rectangle is exactly the same size.

Example 2

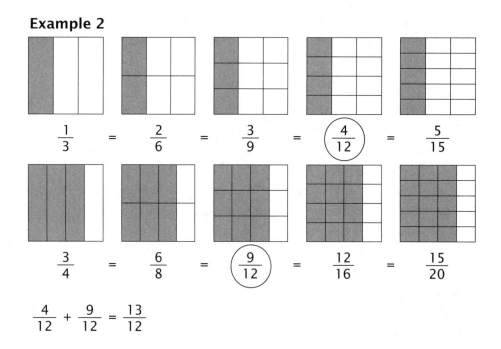

$$\frac{1}{3} = \frac{2}{6} = \frac{3}{9} = \boxed{\frac{4}{12}} = \frac{5}{15}$$

$$\frac{3}{4} = \frac{6}{8} = \boxed{\frac{9}{12}} = \frac{12}{16} = \frac{15}{20}$$

$$\frac{4}{12} + \frac{9}{12} = \frac{13}{12}$$

Notice that 13/12 is greater than 12/12, or one. We will learn more about this kind of fraction later.

After working with the long method for a while, the student should begin to observe a pattern that indicates a shortcut. A quick way to find a common denominator is to crisscross the fraction overlays that represent the denominators.

Example 1 (with the shortcut)

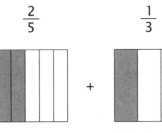

While holding 2/5, take the other third overlay and place it sideways on 2/5.

While holding 1/3, take the other fifth overlay and place it sideways on 1/3.

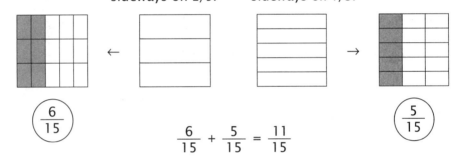

$$\frac{6}{15} + \frac{5}{15} = \frac{11}{15}$$

This quick method will always work. It will not always give you the least common denominator, but it will always give you a common denominator. Don't worry about simplifying fractions for now–that skill will be taught in lesson 12. There will be more about this shortcut in the next lesson.

Example 2 (with the shortcut)

$$\frac{1}{3} \qquad\qquad \frac{3}{4}$$

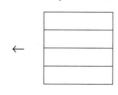 +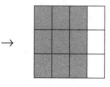

While holding 1/3, take the other fourth overlay and place it sideways on 1/3.

While holding 3/4, take the other third overlay and place it sideways on 3/4.

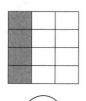 ← ⬜ ⬜ →

$$\left(\frac{4}{12}\right) \qquad\qquad\qquad \left(\frac{9}{12}\right)$$

$$\frac{4}{12} + \frac{9}{12} = \frac{13}{12}$$

The "Rule of Four"
Multi-Step Word Problems

The *Rule of Four* is the Math-U-See name for the shortcut we learned in the last lesson. The drawing below shows 2/5 + 1/3 solved by crisscrossing the overlays. The diagram in Example 1 on the next page uses the "Rule of Four" to solve the same problem without overlays.

Example 1 (with overlays)

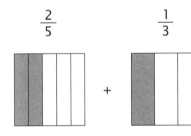

While holding 2/5, take the other third overlay and place it sideways on 2/5.

While holding 1/3, take the other fifth overlay and place it sideways on 1/3.

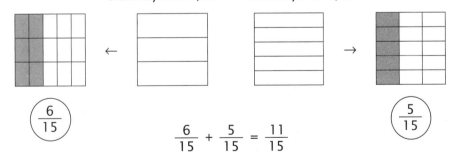

$$\frac{6}{15} + \frac{5}{15} = \frac{11}{15}$$

The Rule of Four is used for adding, subtracting, comparing, and dividing fractions. There are four steps. Steps 1 and 2 show the result of placing the thirds overlay on top of 2/5. The numerator and denominator are each increased by a factor of three. Steps 3 and 4 show the result of placing the fifths overlay on top of 1/3. The numerator and denominator are both increased by a factor of five.

Example 1 (with Rule of Four)

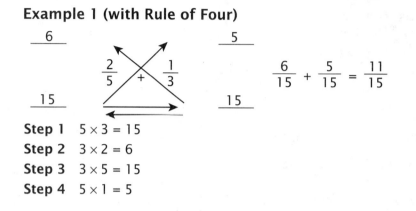

$$\frac{6}{15} + \frac{5}{15} = \frac{11}{15}$$

Step 1 $5 \times 3 = 15$
Step 2 $3 \times 2 = 6$
Step 3 $3 \times 5 = 15$
Step 4 $5 \times 1 = 5$

The Rule of Four always works. It will not always give you the least possible common denominator, but it will always give you some common denominator.

Example 2

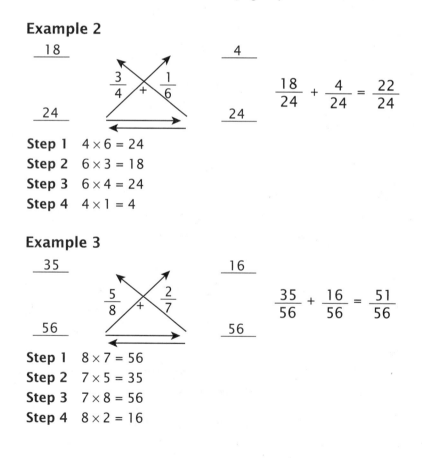

$$\frac{18}{24} + \frac{4}{24} = \frac{22}{24}$$

Step 1 $4 \times 6 = 24$
Step 2 $6 \times 3 = 18$
Step 3 $6 \times 4 = 24$
Step 4 $4 \times 1 = 4$

Example 3

$$\frac{35}{56} + \frac{16}{56} = \frac{51}{56}$$

Step 1 $8 \times 7 = 56$
Step 2 $7 \times 5 = 35$
Step 3 $7 \times 8 = 56$
Step 4 $8 \times 2 = 16$

Multi-Step Word Problems

The student workbook includes some fairly simple two-step word problems. Some students may be ready for more challenging problems. Here are a few to try, along with some tips for solving this kind of problem. You may want to read and discuss these with your student as you work out the solutions together. The purpose is to stretch, not to frustrate. If you do not think the student is ready, you may want to come back to these later.

There are more multi-step word problems in lessons 12, 18, and 24 in this book. The answers are at the end of the solutions section at the back of the book.

1. Tom planted vegetables in a rectangular garden that was 25 feet long and 15 feet wide. He used one third of the area for corn and one fifth of it for peas. How many square feet are left for other vegetables?

Although the problem asks only one question, there are other questions that must be answered first. The key to solving this problem is determining what the unstated questions are. Since the final question is asking for the leftover area, the unstated questions are "What is the total area of Tom's garden?" and "What is the area used for each of the vegetables mentioned?"

You might make a list of questions something like this:

     A. Area of garden in square feet?

     B. Area used for corn?

     C. Area used for peas?

     D. Total area used for peas and corn?

     E. Leftover area?

2. Sarah signed one half of the Christmas cards, and Richard signed three eighths of them. If there are 32 cards in all, how many are left to be signed?

3. Jim wishes to buy three gifts that cost $15.00, $9.00, and $12.00. He has one fourth of the money he needs. How much more money must he earn in order to buy the gifts?

LESSON 7

Comparing Fractions with the Rule of Four

We know that the symbol "=" means "equals" or "is the same as." In *Beta*, we introduced symbols that are used to show that one value is greater than or less than another. As we read an equation from left to right, the symbol ">" means "is greater than," and the symbol "<" means "is less than." These symbols are called *inequality* symbols, and they are used in number sentences called *inequalities*. For example, "nine is greater than three" is written as 9 > 3. "Three is less than nine" is written as 3 < 9. Inequality symbols may also be used to compare fractions.

To remember which symbol is which, some say that the open, or large, end of the symbol always points to the greater number and the small end points to the number that is less. Some students think of the symbol as a hungry alligator with his mouth open, always trying to eat the greater number.

Inequalities with fractions fall into one of two categories: both fractions have the same denominator, or each fraction has a different denominator. If the denominators of two fractions are the same, you simply compare the numerators. It is easy to see that three fourths is greater than one fourth (3/4 > 1/4).

Comparing 2/3 and 3/5 is more difficult. Which is greater? When we added two fractions with different denominators, we first found common denominators (same kind) and then combined the numerators. It the same way, it is easier to compare fractions with different denominators when we first rewrite them with a common denominator. Once the fractions have the same denominator, we compare the numerators. Study the examples on the next page.

Example 1

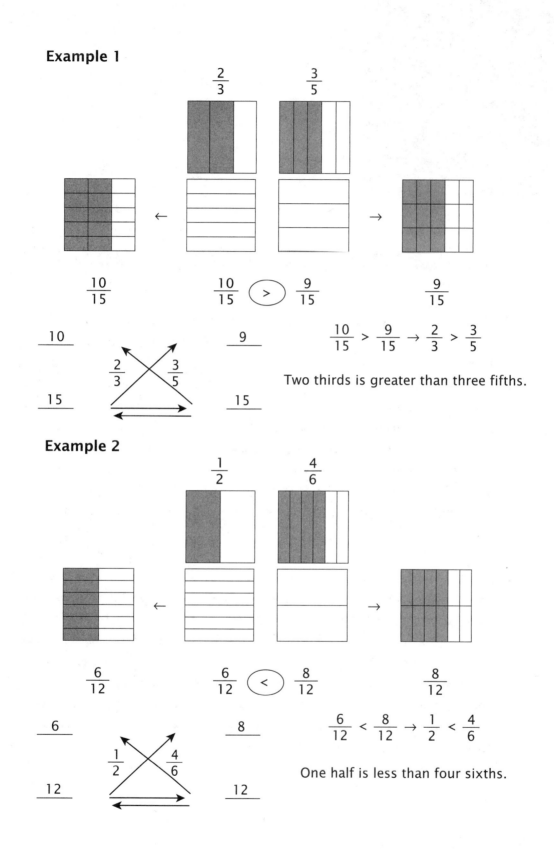

$$\frac{10}{15} > \frac{9}{15} \rightarrow \frac{2}{3} > \frac{3}{5}$$

Two thirds is greater than three fifths.

Example 2

$$\frac{6}{12} < \frac{8}{12} \rightarrow \frac{1}{2} < \frac{4}{6}$$

One half is less than four sixths.

Adding Multiple Fractions

There are several methods that may be used to find the sum of three or more fractions. Let the student become comfortable with the first method before exploring the second and third methods. You may want to spend a day or two on each method to avoid confusion. Examples 1 and 2 use the first method.

Method 1

Simply add two fractions at a time, using the Rule of Four. In Example 1, add 1/2 to 3/5. Add this answer to 4/7.

Example 1

$$\frac{1}{2} + \frac{3}{5} + \frac{4}{7} \rightarrow \frac{1}{2} + \frac{3}{5} = \frac{5}{10} + \frac{6}{10} \rightarrow \frac{11}{10} + \frac{4}{7} = \frac{77}{70} + \frac{40}{70} = \frac{117}{70}$$

Example 2

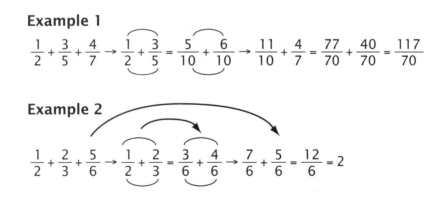

$$\frac{1}{2} + \frac{2}{3} + \frac{5}{6} \rightarrow \frac{1}{2} + \frac{2}{3} = \frac{3}{6} + \frac{4}{6} \rightarrow \frac{7}{6} + \frac{5}{6} = \frac{12}{6} = 2$$

Both examples yield a fraction greater than one. If you remember that the line in a fraction means division, you can divide the numerator by the denominator to get a simpler answer, as in Example 2 above. If you cannot divide evenly, you may write your answer as a whole number and a fraction (mixed number). For now, solutions will be given in unsimplified and simplified forms. Either form is acceptable.

Method 2

This method is a modification of the Rule of Four. Step 1. Multiply the first fraction by the denominators of the second and third fractions. Step 2. Multiply the second fraction by the denominators of the first and third fractions. Step 3. Multiply the third fraction by the denominators of the first and second fractions. Step 4: Add the numerators.

Example 3

$$\frac{1}{2} + \frac{3}{5} + \frac{4}{7} = \frac{1 \times 5 \times 7}{2 \times 5 \times 7} + \frac{3 \times 2 \times 7}{5 \times 2 \times 7} + \frac{4 \times 2 \times 5}{7 \times 2 \times 5} = \frac{35}{70} + \frac{42}{70} + \frac{40}{70} = \frac{117}{70}$$

Example 4

$$\frac{1}{2} + \frac{2}{3} + \frac{5}{6} = \frac{1 \times 3 \times 6}{2 \times 3 \times 6} + \frac{2 \times 2 \times 6}{3 \times 2 \times 6} + \frac{5 \times 2 \times 3}{6 \times 2 \times 3} = \frac{18}{36} + \frac{24}{36} + \frac{30}{36} = \frac{72}{36} = 2$$

Method 3

This is a shorter way for Method 2 that works only if the common denominator for two of the fractions is equal to the denominator of the third fraction. Looking at Example 5, we see that six could be a common denominator because using the Rule of Four on the first two fractions will give a denominator of six for each of them. The third fraction already has a denominator of six. We multiply each of the first two fractions by the number needed to make a denominator of six.

Multiply one half by a factor of three. Multiply two thirds by a factor of two. Five sixths already has a denominator of six, so it is left alone. Once all the denominators are the same, add the numerators.

Example 5

$$\frac{1}{2} + \frac{2}{3} + \frac{5}{6} = \frac{1 \times 3}{2 \times 3} + \frac{2 \times 2}{3 \times 2} + \frac{5}{6} = \frac{3}{6} + \frac{4}{6} + \frac{5}{6} = \frac{12}{6} = 2$$

Method 3 is not a shortcut for Examples 1 and 3 above. No two of the denominators in those problems can be multiplied to equal the third denominator.

Multiplying Fractions
Fraction of a Fraction; Mental Math

We've worked on finding a fraction of one and a fraction of a number; now we'll find a fraction of a fraction. Although the procedure for multiplying fractions is easy to learn, understanding the concept of what it means to multiply by a fraction can be more challenging. An example is "2/5 of 1/3." Notice that we didn't say, "two fifths **times** one third." Even though it is the same operation as multiplying two fractions, I want to relate it to "fraction of a number" and "fraction of one." To stress this relationship, we read it, "two fifths **of** one third."

Similar language is used in most multiplication of fractions word problems. There is a note about word problems on lesson practice 9A in the student workbook. Many of the activities in the application and enrichment pages are designed to aid the student's understanding of word problems involving fractions.

Example 1
Find 2/5 of 1/3, or two fifths of one third.

Step 1
Start with $\dfrac{1}{3}$.

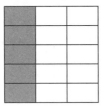

Step 2
Divide it into 5 equal parts.

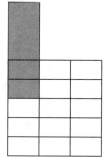

Step 3

Count 2 of those parts.

Pull up the pink insert to do this.

$$\frac{2}{5} \text{ of } \frac{1}{3} = \frac{2}{15}$$

Example 2

Find 2/3 of 5/6, or two thirds of five sixths.

Step 1

Start with $\frac{5}{6}$.

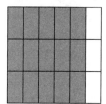

Step 2

Divide into 3 equal parts.

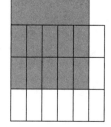

Step 3

Count 2 of those parts.

Pull up the violet insert to do this.

$$\frac{2}{3} \text{ of } \frac{5}{6} = \frac{10}{18}$$

Figure 1 shows another way to understand the formula for a fraction of a fraction (multiplying fractions). It illustrates the final solution for Example 2. To see the factors more clearly, I placed a white piece over the answer so that you can see only the two factors. I hope this helps you "see" numerator times numerator over denominator times denominator.

Figure 1

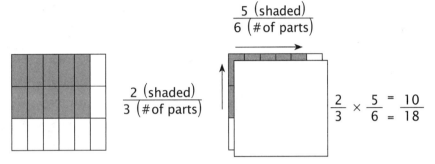

The whole square is an illustration of 2/3 × 5/6, which is a multiplication problem. Within this problem we can see the two partial products. The numerator has the factors 2 × 5. The denominator has the factors 3 × 6. This exercise reveals the formula for multiplying two fractions.

$$\frac{2}{3} \times \frac{5}{6} = \quad \begin{array}{c} = \frac{10}{18} \\ = \frac{10}{18} \end{array}$$

numerator times numerator (shaded rectangle)
denominator times denominator (total number of parts)

Stress that "fraction of a fraction" has the same meaning as "fraction times a fraction." Problems are expressed both ways on the student pages. Some students become confused when multiplying two fractions that are each less than one because the answer is less than the factors. Remind students that taking a fraction of a whole number also results in a part that is smaller than the whole.

Multiplying a Fraction by a Whole Number

Some problems may require multiplying a fraction by a whole number. Remember that a whole number can be written as the whole number of parts over the numeral one. For example, a recipe calls for 1/4 cup of oil. How much oil is needed to make three times that recipe? The solution is $3 \times 1/4$, or $3/1 \times 1/4$, or $3/4$ of a cup of oil.

Mental Math

Here are some more mental math problems for you to read aloud to your student. Try a few at a time, going slowly at first.

1. Four plus five, times two, divided by three, equals ? (6)

2. Thirty-six divided by four, minus two, times five, equals ? (35)

3. Sixty-six divided by six, plus four, divided by three, equals ? (5)

4. Forty-eight divided by eight, times six, plus two, equals ? (38)

5. Six plus seven, minus four, times nine, equals ? (81)

6. Nineteen minus three, divided by four, times seven, equals ? (28)

7. Nine times five, plus three, divided by six, equals ? (8)

8. Twenty-one divided by seven, plus one, times five, equals ? (20)

9. Three times four, divided by two, times nine, equals ? (54)

10. Seven times six, plus two, divided by 11, equals ? (4)

Dividing Fractions with the Rule of Four
Multi-Step Word Problems

I will now introduce what some say is a radical approach to dividing fractions. I tell students to find a common denominator and then divide the numerators. Students are usually told to invert the second fraction and multiply, but this is hard to explain or understand. My method, while not as quick as inverting and multiplying, is much easier to understand and relates to what a student already knows. The shortcut of inverting and multiplying (multiplying by the reciprocal) will be taught in lesson 23, when students are more experienced with fractions.

Students know that to add or subtract fractions, they must find a common denominator and then add or subtract the numerators. We will use the same strategy for division of fractions. Word the question in the same way you did for division of whole numbers. When we taught $6 \div 2$, we said, "How many twos can we count out of six?" The answer is three. The same question applies to dividing fractions. For $4/5 \div 1/5$, take $1/5$ in one hand and $4/5$ in the other hand and ask, "How many one fifths can I count out of four fifths?" You can clearly see that the answer is four (Example 1). For $3/4 \div 1/4$, ask, "How many one fourths can I count out of three fourths?" The answer is three (Example 2 on the next page).

Example 1

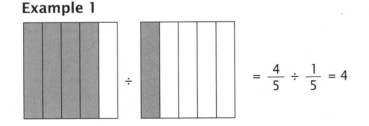

$$= \frac{4}{5} \div \frac{1}{5} = 4$$

Example 2

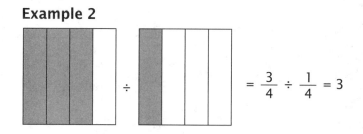

$$= \frac{3}{4} \div \frac{1}{4} = 3$$

Because the denominators of the fractions are the same, dividing them always yields one. Remember that anything divided by one is the number itself and does not affect the answer.

Example 3

$$\frac{4}{5} \div \frac{1}{5} = \frac{4}{1} = 4$$

A *unit fraction* is a fraction with a numerator of one. In Example 3, the fraction 4/5 is being divided by the unit fraction 1/5. You may want to have the student practice dividing by a unit fraction first, as it may be easier to visualize the process.

Dividing Fractions and Whole Numbers

"How many quarters are in one dollar?" The equation is 1 ÷ 1/4. We can't figure this until we give the parts the same denominator. Changing 1 to 4/4, we have 4/4 ÷ 1/4. Think, "How many one fourths can we count out of four fourths?" The answer is four fourths, or four quarters.

"We have 1/3 of a pie. If we divide the pie between two people, what part of a pie will each get?" The expression is 1/3 ÷ 2, or 1/3 ÷ 2/1. Using the Rule of Four, we can rewrite it as 1/3 ÷ 6/3. The answer is 1 ÷ 6, or 1/6 of a pie for each person.

Division of Fractions with Different Denominators

Example 4

Solve: 4/5 divided by 1/3

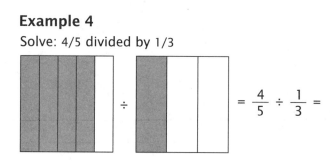

$$= \frac{4}{5} \div \frac{1}{3} =$$

Notice how difficult it is to estimate in this form.

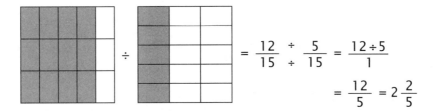

$$= \frac{12}{15} \div \frac{5}{15} = \frac{12 \div 5}{1}$$

$$= \frac{12}{5} = 2\frac{2}{5}$$

12/5 is the same as 12 ÷ 5, or 2⅖. Write your answer in either form.

The original question is "How many one thirds can I count out of four fifths?" After finding a common denominator, the question is, "How many five fifteenths can I count out of twelve fifteenths?" Look at the picture and ask, "How many groups of five can I count out of 12?" The answer is two groups of five with two of the fifths remaining, or 2⅖.

Example 5
Solve: 2/3 divided by 1/4

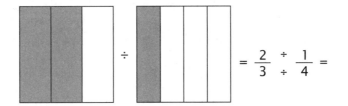

$$= \frac{2}{3} \div \frac{1}{4} =$$

Notice how difficult it is to estimate in this form.

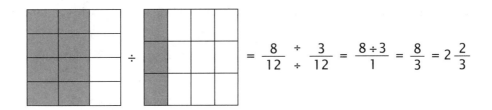

$$= \frac{8}{12} \div \frac{3}{12} = \frac{8 \div 3}{1} = \frac{8}{3} = 2\frac{2}{3}$$

8/3 is the same as 8 ÷ 3, or 2⅔. Write your answer in either form.

Look at the picture and ask, "How many groups of three can I count out of eight?" The answer is 8/3, or 2⅔.

Example 6
Solve: 1/4 divided by 3/5

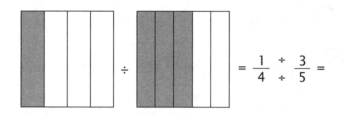

$= \dfrac{1}{4} \div \dfrac{3}{5} =$

The question is "How many three fifths are there in one fourth?"

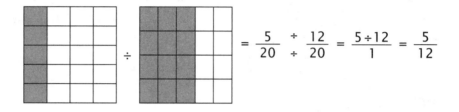

$= \dfrac{5}{20} \div \dfrac{12}{20} = \dfrac{5 \div 12}{1} = \dfrac{5}{12}$

Remember that a fraction is another way of writing a division problem. In this case, we are dividing a lesser number by a greater, so the answer is less than one. We leave the answer written as a fraction.

Fraction Word Problems

Have students carefully observe the language used in the division word problems on the lesson practice pages. Rather than just looking for clue words, have them read carefully and try to visualize what is being asked in each question. Systematic review pages 10D and 10E have specially-marked word problems to illustrate the four basic operations using fractions.

Finding Common Factors and Divisibility

Factoring is the opposite of multiplying. In a multiplication problem, we are given two factors, and we build a rectangle to find the area. In *factoring*, we are given the area and build a rectangle with that area to find the pairs of factors.

Example 1
Find the factors of 6.

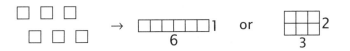

Building shows us two pairs of factors. They are 1 × 6 and 2 × 3. We can list the factors as 1, 2, 3, and 6.

Example 2
Find the factors of 12.

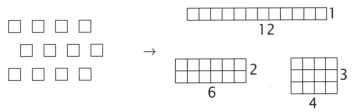

Building shows us three pairs of factors. They are 1 × 12, 2 × 6, and 3 × 4. We can list the factors as 1, 2, 3, 4, 6, and 12.

Divisibility

You can tell a lot about a number by looking at the units place. If a number ends in 2, 4, 6, 8, or 0, it is an even number. *Divisibility* refers to being able to divide by a number without getting a remainder. Divisibility clues us in to factors. For example, even numbers are always divisible by two. This means that, since you can always divide an even number by two with no remainder, at least one of the factors of an even number is two.

2	12
4	14
6	16
8	18
10	20

If a number ends in zero, it is a multiple of 10. A *multiple* is the result of multiplying by some number. If a number is a multiple of 10, it means that at least one of its factors is 10. Patterns emerge when you look at the skip counting facts. Look at the 10 facts below and notice that each number ends in zero and is a multiple of 10.

10	60
20	70
30	80
40	90
50	100

All the multiples of five have a five or a zero in the units place. Therefore, we can conclude that, if a number ends in five or zero, it is divisible by five, and at least one of its factors is five.

5	30
10	35
15	40
20	45
25	50

Multiples of two, five, and ten are easily recognized by looking at the digit in the units place. A different approach is needed to find multiples of three and nine.

Look at the answers to the three facts shown below. When you add up the digits in each number, the sum is either three or a multiple of three every time. To tell whether a number is divisible by three, simply add up the digits. This works even for greater numbers. If a number's digits add up to three or a multiple of three, then the number is divisible by three, and at least one of its factors is three.

3		18	1 + 8 = 9
6		21	2 + 1 = 3
9		24	2 + 4 = 6
12	1 + 2 = 3	27	2 + 7 = 9
15	1 + 5 = 6	30	3 + 0 = 3

Example 3
Is 4,715 a multiple of 3?

4 + 7 + 1 + 5 = 17.

17 is not a multiple of 3, so 4,715 is not a multiple of 3.

Example 4
Is 1,218 a multiple of 3?

1 + 2 + 1 + 8 = 12.

12 is a multiple of 3, so 1,218 is a multiple of 3.

To tell if a number is divisible by nine, simply add up the digits. If the digits add up to nine or a multiple of nine, the number is divisible by nine, and at least one of its factors is nine.

9	0 + 9 = 9	54	5 + 4 = 9
18	1 + 8 = 9	63	6 + 3 = 9
27	2 + 7 = 9	72	7 + 2 = 9
36	3 + 6 = 9	81	8 + 1 = 9
45	4 + 5 = 9	90	9 + 0 = 9

Example 5

Is 7,218 a multiple of 9?

7 + 2 + 1 + 8 = 18.

18 is a multiple of 9, so 7,218 is a multiple of 9.

Example 6

Is 8,729 a multiple of 9?

8 + 7 + 2 + 9 = 26.

26 is not a multiple of 9, so 8,729 is not a multiple of 9.

There are some rather complex methods for testing to see if a number is divisible by four, six, or eight, but these methods add needless work to the process. If a number is a multiple of four, we can divide by two twice. A number divisible by eight can be divided by two three times.

If a number is a multiple of six, then it must also be a multiple of two and three. We can divide by two and then by three. Knowing how to recognize the multiples of 2, 3, 5, 9, and 10 is very helpful when simplifying fractions, which will be discussed in lesson 12.

Greatest Common Factor

There are three important parts of the term *"greatest common factor"*: the *factor*, the *common* factor, and the *greatest* common factor. The factors of six are 1×6 and 2×3 and may be written in order as 1, 2, 3, and 6. The factors of 15 are 1×15 and 3×5 and may be written in order as 1, 3, 5, and 15. The common factors of 6 and 15 are 1 and 3. These common factors are underlined in Figure 1. The *greatest* common factor (GCF), or the greatest of the common factors, is 3.

Figure 1

6	1, 2, 3, 6
15	1, 3, 5, 15

Example 7
Find the GCF of 12 and 18.

1. The factors of 12 and 18 are as follows:

    12   1, 2, 3, 4, 6, 12
    18   1, 2, 3, 6, 9, 18

2. The common factors of 12 and 18 are underlined.

    12   <u>1</u>, <u>2</u>, <u>3</u>, 4, <u>6</u>, 12
    18   <u>1</u>, <u>2</u>, <u>3</u>, <u>6</u>, 9, 18

3. The factors 1, 2, 3, and 6 are each common factors of 12 and 18, but 6 is the greatest common factor.

Example 8
Find the GCF of 18 and 27.

1. The factors of 18 and 27 are as follows:

    18   1, 2, 3, 6, 9, 18
    27   1, 3, 9, 27

2. The common factors of 18 and 27 are underlined.

    18   <u>1</u>, 2, <u>3</u>, 6, <u>9</u>, 18
    27   <u>1</u>, <u>3</u>, <u>9</u>, 27

3. The factors 1, 3, and 9 are each common factors of 18 and 27, but 9 is the greatest common factor.

Example 9

Find the GCF of 30 and 45.

1. The factors of 30 and 45 are as follows:

     30   1, 2, 3, 5, 6, 10, 15, 30
     45   1, 3, 5, 9, 15, 45

2. The common factors of 30 and 45 are underlined.

     30   <u>1</u>, 2, <u>3</u>, <u>5</u>, 6, 10, <u>15</u>, 30
     45   <u>1</u>, <u>3</u>, <u>5</u>, 9, <u>15</u>, 45

3. The factors 1, 3, 5, and 15 are each common factors of 30 and 45, but 15 is the greatest common factor.

Simplifying Fractions—Common Factors
Word Problems

Simplifying fractions is the opposite of making equivalent fractions with the overlays. When making equivalent fractions, we build a fraction and then place overlays horizontally on top to change the number of pieces without affecting the amount. Remember, we have same amount but more pieces. Let's focus on what happens when we place the overlay on top of the fraction we built.

Example 1

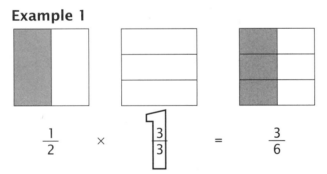

$$\frac{1}{2} \qquad \times \qquad \frac{3}{3} \qquad = \qquad \frac{3}{6}$$

Beginning with 1/2, the numerator and denominator are each increased by a common factor of three. Notice that we are actually multiplying both the numerator and the denominator by three.

$$\frac{1}{2} = \frac{1 \times 3}{2 \times 3} = \frac{3}{6}$$

In Example 1 we are placing only one overlay on top, but the numerator and denominator are each being increased by the common factor of three.

Another way to look at this is to notice the over factor and the up factor. In Example 2 on the next page, the diagram from Example 1 is covered to reveal the edges, or factors. The factors are over 1/2 and up 3/3. Emphasize that 3/3 is equivalent to one. We still have one whole, but it has been cut into three pieces. Using the overlays in this way is like multiplying by one.

Example 2

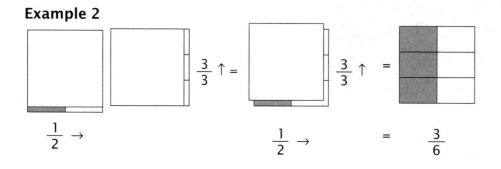

$$\frac{1}{2} \rightarrow \qquad\qquad \frac{1}{2} \rightarrow \qquad = \qquad \frac{3}{6}$$

Example 3

Put overlay on.

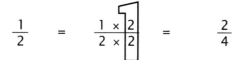

$$\frac{1}{2} \quad = \quad \frac{1 \times \boxed{2}}{2 \times \boxed{2}} \quad = \quad \frac{2}{4}$$

Starting with 1/2, the numerator and denominator are each increased by a common factor of two.

Example 4

Put overlay on.

$$\frac{1}{2} \quad = \quad \frac{1 \times \boxed{4}}{2 \times \boxed{4}} \quad = \quad \frac{4}{8}$$

Starting with 1/2, the numerator and denominator are each increased by a common factor of four.

Example 5 (reversing Example 1)

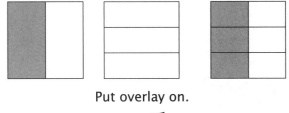

Put overlay on.

$$\frac{1}{2} \quad = \quad \frac{1 \times \boxed{3}}{2 \times \boxed{3}} \quad = \quad \frac{3}{6}$$

This is the problem from Example 1. Notice that we *increased* by a common factor of three when we *multiplied* both the numerator and denominator by 3.

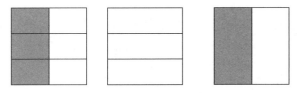

Take overlay off.

$$\frac{3}{6} \quad = \quad \frac{3 \div \boxed{3}}{6 \div \boxed{3}} \quad = \quad \frac{1}{2}$$

Here is the reverse of the same example. Starting with 3/6, we *divided* both the numerator and denominator by 3 and *simplfied* the fraction.

Example 6 (reversing Example 3)

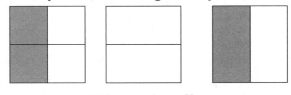

Take overlay off.

$$\frac{2}{4} \quad = \quad \frac{2 \div \boxed{2}}{4 \div \boxed{2}} \quad = \quad \frac{1}{2}$$

Starting with 2/4, the numerator and denominator are each divided by a common factor of two. The simplified fraction is 1/2.

Example 7 (reversing Example 4)

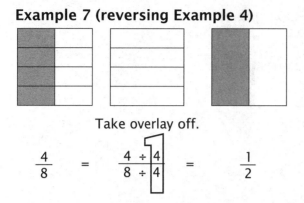

Take overlay off.

$$\frac{4}{8} = \frac{4 \div 4}{8 \div 4} = \frac{1}{2}$$

Starting with 4/8, the numerator and denominator are each divided by a common factor of four. The simplified fraction is 1/2.

Remember what we observed in Example 2. It is the up factor that either increases or simplifies a fraction. Looking at the up factor tells us the common factor by which the fraction is being increased or simplified.

Word Problems

Here are a few more multi-step word problems to try. You may want to read and discuss these with your student as you work out the solutions together. Again, the purpose is to stretch, not to frustrate. If you do not think the student is ready, you may want to come back to these later.

The answers are at the end of the solutions section at the back of this book.

1. Jennifer had $30 to spend on herself. She spent one fifth of the money on a sandwich, one sixth for a ticket to a museum, and one half of it on a book. How much money does Jennifer have left over?

2. Mark drove for one half of the trip to the mountains, and Justin drove for one fourth of the trip. Gina and Kaitlyn divided the rest of the driving evenly between them. If the entire trip covered 128 miles, how far did Kaitlyn drive?

3. Two pie pans of the same size remain on the counter. One pan has one half of a pie, and the other has three fourths of a pie left in it. Mom wishes to divide all the pie into pieces that are each one eighth of a pie. How many pieces of pie will she have when she is finished?

Simplifying Fractions—Prime Factors

A number for which we can build more than one rectangle is called a *composite number*. The composite numbers between 1 and 24 are 4, 6, 8, 9, 10, 12, 14, 15, 16, 18, 20, 21, 22, and 24. A number for which we can build only one rectangle, such as five or seven, is called a prime number. A *prime number* is a whole number greater than one which has exactly two factors: one and itself.

To illustrate this concept, I give students seven green blocks and offer a reward for the first person who can make two different rectangles using all of the blocks in each rectangle. They soon find out that only one rectangle can be built. (A rectangle that is one by seven is the same as a rectangle that is seven by one, except that one is standing, and the other is reclining.) I then say, "What do we call a number for which you can build only one rectangle?" The answer is, "A prime number." The definition given above makes sense once the student has experimented with the manipulatives.

Example 1
Find all the possible factors of seven and tell whether the number is prime or composite.

$$\underbrace{\boxed{\,|\,\,|\,\,|\,\,|\,\,|\,\,|\,}}_{7}\ 1 \longrightarrow$$ The factors are 7×1 or 1×7, so 7 is a prime number.

The prime numbers between 1 and 24 are 2, 3, 5, 7, 11, 13, 17, 19, and 23. Remember that the number one is not considered a prime number. Ask the students to build any of the numbers listed above, and they will find that there is only one rectangle to be built. A prime number has only one set of factors.

Finding the prime factors of a number is called *prime factorization.* Using a factor tree or repeated division can make the process easier. Here is an example of each method used to find the prime factors of 12 and 18.

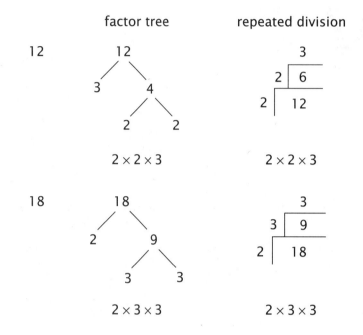

Example 2

Simplify 12/18 using prime factorization.

We found the prime factors of 12 and 18 with the factor tree and with repeated division. Now we have to find which numbers in the numerator and denominator may be divided to make one.

$$\frac{12}{18} = \frac{2 \times 2 \times 3}{2 \times 3 \times 3} = \frac{2}{2} \times \frac{2}{3} \times \frac{3}{3} = 1 \times 1 \times \frac{2}{3}$$

Notice the outline of a one around 2/2 and 3/3. The fraction 2/2 (two divided by two) equals one, and the fraction 3/3 also equals one. Some refer to this process of making ones as "canceling," or making the twos and threes disappear. The numbers don't really cancel or disappear; instead, because they equal one, they do not change the value of 2/3.

The prime factorization method of simplifying fractions is attractive because we do not have to recognize the greatest common factor (GCF). In some cases, the GCF is very obvious, so you should use it instead to simplify. When it isn't so obvious, simply find the prime factors and start dividing until the fraction is in its simplest form and no more ones may be divided out of it.

Example 3

Simplify 10/14 using prime factorization.

First find all the prime factors of 10 and 14, using either a factor tree or repeated division.

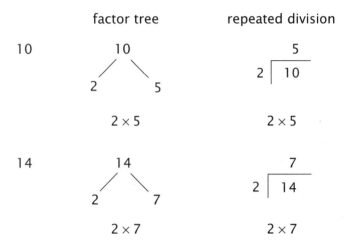

Now find the numbers in the numerator and denominator that may be divided to make one.

$$\frac{10}{14} = \frac{2 \times 5}{2 \times 7} = \frac{2}{2} \times \frac{5}{7} = 1 \times \frac{5}{7}$$

Two divided by two is one.
One times five sevenths is five sevenths.
Thus, ten fourteenths simplifies to five sevenths.

LESSON 14

Linear Measure

Recall how we calculated a fraction of a number and a fraction of one in previous lessons. We first looked at the starting value, divided it into equal parts (denominator), and finally counted the number of equal parts (numerator). In this lesson, we will be finding the fraction of an inch. Let's start with 3/4 of one inch. Start with one inch └─────────┘, divide it into four equal parts └─┴─┴─┴─┘, and count three of those parts └▨▨▨┴─┘.

Throughout this book, fractions have been taught using the method of counting the total spaces to find the denominator and then counting a certain number of those spaces for the numerator. By using this approach, the student will be able to read any ruler, even ones with less-common divisions such as fifths and tenths, instead of memorizing only the common halves, fourths, and eighths.

Here are examples of some "unusual" measurement problems to make sure the concept is mastered. When working through Examples 1 and 2, use the overlays to create the fraction, pretending the width of the overlay represents one inch instead of five inches.

Example 1
Draw a line that is 4/5 of an inch long, or 4/5".

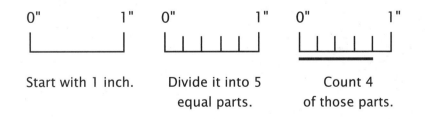

Start with 1 inch.   Divide it into 5 equal parts.   Count 4 of those parts.

Example 2

Draw a line 2/3 of an inch, or 2/3".

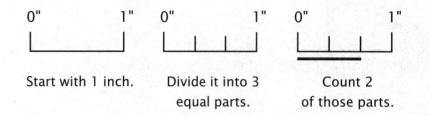

Start with 1 inch.   Divide it into 3 equal parts.   Count 2 of those parts.

Example 3

How long is the line segment?

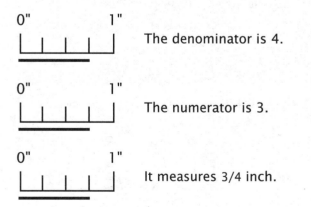

First find the denominator by counting the number of spaces. There are 6 spaces.

Find the numerator by counting the length of the line. It is 4 spaces long.

The line is 4/6 of an inch, or 4/6" long.

Example 4

How long is the line segment?

The denominator is 4.

The numerator is 3.

It measures 3/4 inch.

In the fraction kit, there are two 1/2" × 5" clear overlays used to create eighths and sixteenths, which are commonly used in measurement.

Figure 1
Place the 1/8 overlay on top of 1/2 to show that 1/2 equals 4/8.

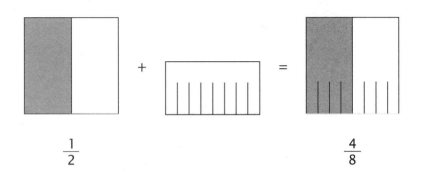

$$\frac{1}{2} \qquad\qquad\qquad \frac{4}{8}$$

Figure 2
Place the 1/16 overlay on top of 1/2 to show that 1/2 equals 8/16.

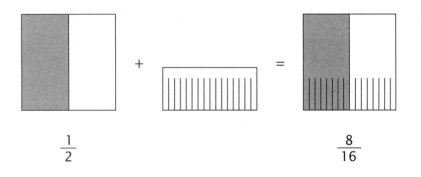

$$\frac{1}{2} \qquad\qquad\qquad \frac{8}{16}$$

Figure 3
To show that 1/2 is equal to 2/4, use the 1/4 overlay.

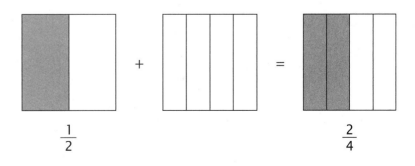

$$\frac{1}{2} \qquad\qquad\qquad \frac{2}{4}$$

Example 5
Show that 3/4 = 6/8 = 12/16.

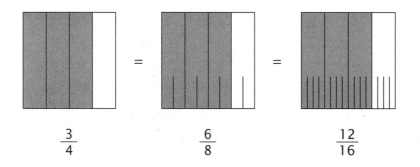

$$\frac{3}{4} \qquad \frac{6}{8} \qquad \frac{12}{16}$$

Now that we have made eighths and sixteenths, notice how simplifying fractions applies to measurement. The measurements 8/16″, 4/8″, and 2/4″ may all be expressed as 1/2″. This concept can be illustrated with open-end wrenches. There are 1/2″ wrenches, 5/8″ wrenches, and 3/4″ wrenches, to name a few. There are no 2/8″, 4/8″, and 6/8″ wrenches or 8/16″, 10/16″, and 12/16″ wrenches because we always simplify when giving a measure.

A succession of rulers is shown below (not to scale). Observe how the last ruler is a compilation of the first four rulers. There are different types of rulers, but the last one is most commonly used in the United States.

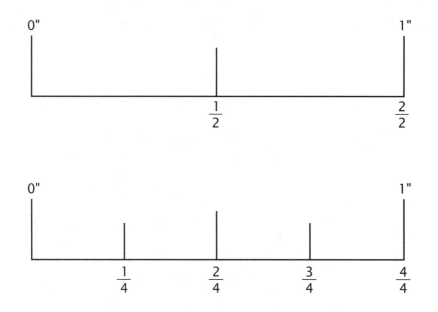

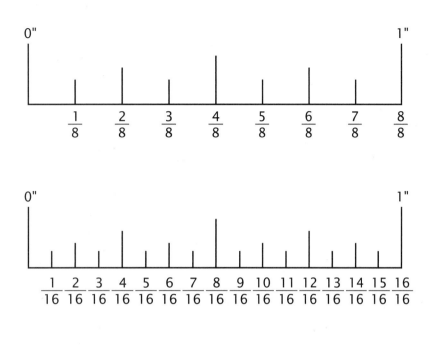

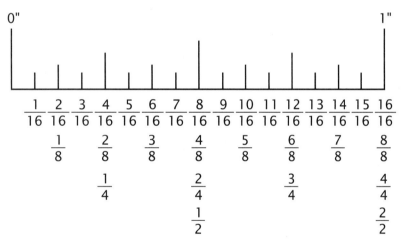

Mixed Numbers and Improper Fractions
Mental Math

A *mixed number* is a combination of a whole number and a fraction. Perhaps mixed numbers should be called frac-bers (*frac* from the fraction and *ber* from the whole number), or num-tions (*num* from the whole number and *tion* from the fraction). The combination 2½ is called a mixed number; 2 is the whole number, and ½ is the fraction.

When a fraction has a numerator that is less than the denominator, it is called a *proper fraction.* When the numerator is equal to the denominator, the fraction is equal to one. When the numerator is greater than the denominator, the fraction is called an *improper fraction.* I mention this now because when you change a mixed number to a fraction, the resulting fraction will always be an improper fraction.

To change a mixed number to an improper fraction, use the colored pieces from the overlay set that represent one. These include the large white or green pieces, the pink 3/3 pieces, the blue 5/5 piece, or one of the other colored pieces that represent one.

Some students may be helped to understand converting a mixed number to an improper fraction by considering money. If you have two dollars and a quarter, how many quarters do you have? The answer is four quarters for each dollar, plus one quarter, or nine quarters altogether. See Example 1. The expression 2¼ is a mixed number, and 9/4 is an improper fraction.

The process of breaking a number down into parts is sometimes called *decomposition.* In this case, the parts of the mixed number are added together to make an improper fraction with the same value as the original number.

Example 1

$$2\frac{1}{4} = 1 + 1 + \frac{1}{4} = \frac{4}{4} + \frac{4}{4} + \frac{1}{4} = \frac{9}{4}$$

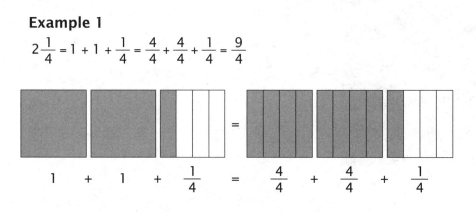

$$1 \quad + \quad 1 \quad + \quad \frac{1}{4} \quad = \quad \frac{4}{4} \quad + \quad \frac{4}{4} \quad + \quad \frac{1}{4}$$

Example 2

Change 1⅗ to an improper fraction.

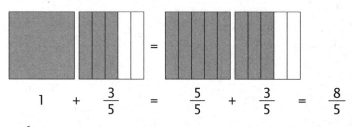

$$1 \quad + \quad \frac{3}{5} \quad = \quad \frac{5}{5} \quad + \quad \frac{3}{5} \quad = \quad \frac{8}{5}$$

1⅗ is a mixed number, and 8/5 is an improper fraction.

To change an improper fraction to a mixed number, do the opposite. I find that thinking about money and quarters is very helpful in understanding this process. Ask the student, "If you have nine quarters, how many dollars do you have?" The answer is, "There are four quarters in a dollar, so eight quarters is two dollars, with one quarter is left over." The overlays show this in Example 3.

Example 3

Change 9/4 to a mixed number.

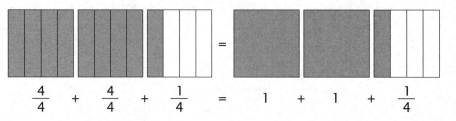

$$\frac{4}{4} \quad + \quad \frac{4}{4} \quad + \quad \frac{1}{4} \quad = \quad 1 \quad + \quad 1 \quad + \quad \frac{1}{4}$$

Start with 9/4 and subtract 4/4 (which is one) to get 5/4. Then subtract 4/4 again (another one), and 1/4 is left. Therefore, 9/4 is the same as 2 and 1/4.

Example 4

Change 12/5 to a mixed number.

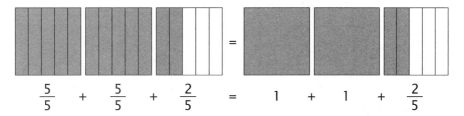

$$\frac{5}{5} \ + \ \frac{5}{5} \ + \ \frac{2}{5} \ = \ 1 \ + \ 1 \ + \ \frac{2}{5}$$

Start with 12/5 and subtract 5/5 (which is one) to get 7/5. Then subtract 5/5 again (another one), and 2/5 is left. Therefore, 12/5 is the same as 2 and 2/5.

Example 5

Change 5/3 to a mixed number.

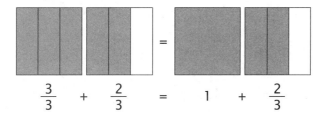

$$\frac{3}{3} \ + \ \frac{2}{3} \ = \ 1 \ + \ \frac{2}{3}$$

Start with 5/3 and subtract one in the form of 3/3 to get 2/3. Therefore, 5/3 is the same as 1 and 2/3.

Exponents

The student book has been reviewing area. Be sure to teach the Quick Tip on 15E. It shows how to use an *exponent* to indicate that a number is to be multiplied by itself. An exponent may also be used to label the answer to an area problem. Exponents will be taught in more detail in *Zeta*.

Mental Math

Here are some more mental math problems for you to read aloud to your student. Read the problems in the order given.

1. Twenty-four divided by four, plus two, times six, equals ? (48)

2. Fifteen minus eight, times two, plus five, equals ? (19)

3. Three plus three, times five, divided by 10, equals ? (3)

4. Twenty-five divided by five, times seven, plus four, equals ? (39)

5. Fifty minus one, divided by seven, plus two, equals ? (9)

6. Sixty-four divided by eight, minus two, times four, equals ? (24)

7. Twenty-five plus two, divided by three, times six, equals ? (54)

8. Three times five, plus five, divided by two, equals ? (10)

9. Fifty-six divided by seven, divided by two, times seven, equals ? (28)

10. Nine times nine, minus one, divided by 10, equals ? (8)

Linear Measure with Mixed Numbers

Now we will combine what we know about linear measure with what we know about mixed numbers. In linear measure we learned how to read a fraction of an inch. A mixed number employs a whole number along with a fraction of a number. To read a ruler, start at the left and read the number of full inches. Continue looking to the right and read the fraction of the next inch. Combining the whole number of inches with the fraction of an inch produces the correct measure.

Example 1
How long is the line?

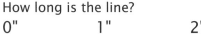

Starting at the left, we see two full inches.
Beginning at the two-inch mark, we read a fraction of an inch, which is 1/4 inches. Putting them together, we see the line is 2¼ inches long.

Example 2
How long is the line?

Starting at the left, we see one full inch.
Beginning at the one-inch mark, we read a fraction of an inch, which is 9/16 inches. Putting them together, we see the line is 1⁹⁄₁₆ inches long.

Example 3

How long is the line? Be careful; this is not a standard ruler.

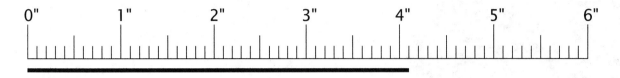

Starting at the left, we see four full inches. Beginning at the four-inch mark, we read a fraction of an inch, which is 3/10 inches. Putting them together, we see the line is $4\frac{3}{10}$ inches long.

Example 4

How long is the line?

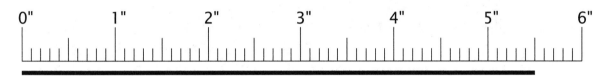

Starting at the left, we see five full inches.
Beginning at the five-inch mark, we read a fraction of an inch. This is 5/10 inch, which simplifies to 1/2 inch.

Putting them together, we see the line is $5\frac{1}{2}$ inches long.

Addition of Mixed Numbers

With Regrouping; Word Problems

Adding mixed numbers with regrouping is very similar to regrouping with whole numbers. Instead of converting 10 units to one ten, we are rewriting a fraction as an equivalent whole number. In Example 1, we rewrite the fraction 3/3 as the whole number one.

Working through the problem, 2 + 1 = 3, and 2/3 + 2/3 = 4/3. However, 4/3 is an improper fraction. We can rewrite 4/3 as 3/3 plus 1/3. Since 3/3 is equal to one, 4/3 is the same as 1⅓. Adding 3 and 1⅓ gives us 4⅓ as our final answer.

Example 1

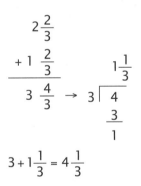

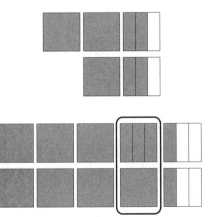

$$3 + 1\frac{1}{3} = 4\frac{1}{3}$$

Example 2

$$1\frac{4}{5}$$
$$+\ 2\frac{3}{5}$$
$$\overline{\ \ 3\frac{7}{5}\ } \rightarrow 5\overline{\smash{)}\ \begin{array}{c}1\frac{2}{5}\\ 7\\ \underline{5}\\ 2\end{array}}$$

$$3 + 1\frac{2}{5} = 4\frac{2}{5}$$

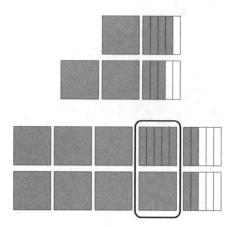

Example 3

$$2\frac{3}{4}$$
$$+\ \ \frac{3}{4}$$
$$\overline{\ \ 2\frac{6}{4}\ } \rightarrow 4\overline{\smash{)}\ \begin{array}{c}1\frac{2}{4}\\ 6\\ \underline{4}\\ 2\end{array}}$$

$$2 + 1\frac{2}{4} = 3\frac{2}{4} = 3\frac{1}{2}$$

Word Problems

Here are a few more multi-step word problems to try. You may want to read and discuss these with your student as you work out the solutions together. Again, the purpose is to stretch rather than frustrate. If you do not think the student is ready, you may want to come back to these later.

The answers are at the end of the solutions section at the back of this book.

1. A rectangular field is 63 yards long and 21 yards wide. A fence is needed for the perimeter of the field. Fencing is also needed to divide the field into three square sections. How many feet of fencing are needed? (It is a good idea to make a drawing for this one.)

2. One half of the people at the game wore their team colors. Two thirds of those people wore team hats as well. One fourth of those with team colors and team hats had banners to wave. Twenty-five people had team colors and banners but not hats. One hundred people had only banners. If there were 1,824 people at the game, how many had banners?

3. A cube-shaped pool is half full of water. If the water is 36 inches deep, how much would the water in the pool weigh if the pool were filled to the brim? (One cubic foot of water weighs 63 pounds.)

Subtraction of Mixed Numbers
With Regrouping

Now that we have learned how to add mixed numbers, we can start subtracting them. Often we will need to regroup before we subtract. In Example 1, we can't subtract 2/3 from 1/3, so we need to regroup one of the four, leaving three. Then we change 1 to 3/3 so that we can combine it with 1/3 to make 4/3. Finally, we subtract the whole numbers and the fractions.

Example 1

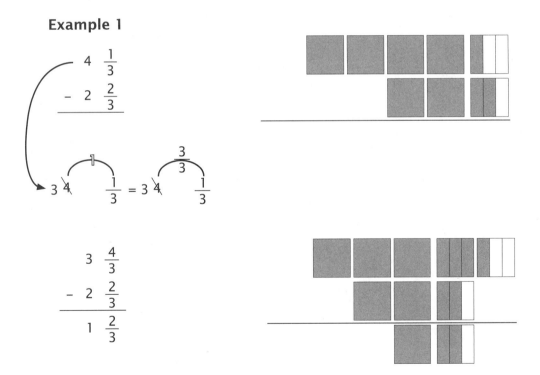

$$4 \frac{1}{3}$$
$$- 2 \frac{2}{3}$$

$$3 \,\, 4 \quad \frac{1}{3} = 3 \,\, 4 \quad \frac{1}{3}$$

$$3 \frac{4}{3}$$
$$- 2 \frac{2}{3}$$
$$1 \frac{2}{3}$$

Example 2

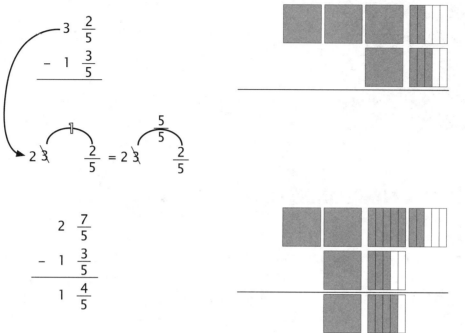

$$3 \frac{2}{5}$$
$$- 1 \frac{3}{5}$$

$$2 \overset{\cancel{3}}{} \overset{1}{\frown} \frac{2}{5} = 2 \overset{\cancel{3}}{} \overset{\frac{5}{5}}{\frown} \frac{2}{5}$$

$$2 \frac{7}{5}$$
$$- 1 \frac{3}{5}$$
$$1 \frac{4}{5}$$

Subtraction of Mixed Numbers
Using the "Same Difference Theorem"

Be sure the student has mastered the material taught in lesson 19 before attempting this lesson. After teaching the "same difference theorem", have the student use this method to redo some of the problems from the last lesson and compare their answers to the original answers.

The *"same difference theorem"* is a Math-U-See idea. It states that we can add the same amount to both parts of a subtraction problem without changing the difference between the two parts of the problem.

Example 1 (from Example 1, lesson 19)

$$
\begin{array}{rcl}
4\dfrac{1}{3} + \dfrac{1}{3} & = & 4\dfrac{2}{3} \\[2mm]
-\,2\dfrac{2}{3} + \dfrac{1}{3} & = & -\,3 \\[2mm]
\hline
& & 1\dfrac{2}{3}
\end{array}
$$

We ask, "What do I have to add to 2/3 to make it a whole number?"
The answer is 1/3.

Now we have to add 1/3 to $4\frac{1}{3}$ as well, which makes $4\frac{2}{3}$.

The problem is now $4\frac{2}{3}$ minus 3, which is easier to subtract.

Example 2 (from Example 2, lesson 19)

$$3\frac{2}{5} + \frac{2}{5} = \quad 3\frac{4}{5}$$
$$-\,1\frac{3}{5} + \frac{2}{5} = \quad -\,2$$
$$\overline{\qquad\qquad} \quad \overline{\quad 1\frac{4}{5}\quad}$$

We ask, "What do I have to add to 3/5 to make it a whole number?"
The answer is 2/5.

Now we have to add 2/5 to $3\frac{2}{5}$ as well, which makes $3\frac{4}{5}$.

The problem is now $3\frac{4}{5}$ minus 2, which is easier to subtract.

Adding Mixed Numbers
Regrouping and Unequal Denominators; Mental Math

In this lesson we will learn how to add mixed numbers when the fractions have different denominators. In the examples, we separate the fractions, find a common denominator, rewrite the problem, and then perform the appropriate operation, as in previous lessons.

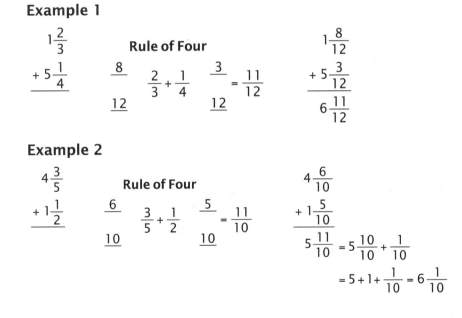

Example 1

$$1\frac{2}{3}$$
$$+\ 5\frac{1}{4}$$

Rule of Four

$$\frac{8}{12} \qquad \frac{2}{3}+\frac{1}{4} \qquad \frac{3}{12} = \frac{11}{12}$$

$$1\frac{8}{12}$$
$$+\ 5\frac{3}{12}$$
$$6\frac{11}{12}$$

Example 2

$$4\frac{3}{5}$$
$$+\ 1\frac{1}{2}$$

Rule of Four

$$\frac{6}{10} \qquad \frac{3}{5}+\frac{1}{2} \qquad \frac{5}{10} = \frac{11}{10}$$

$$4\frac{6}{10}$$
$$+\ 1\frac{5}{10}$$
$$5\frac{11}{10} = 5\frac{10}{10}+\frac{1}{10}$$
$$= 5+1+\frac{1}{10} = 6\frac{1}{10}$$

Example 3

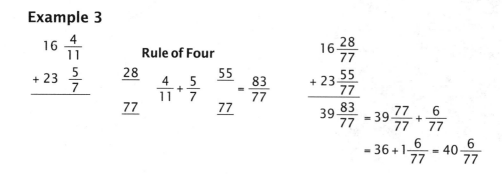

Mental Math

Read these aloud to your student and have him solve mentally.

1. Thirty-six divided by six, plus three, times eight, equals ? (72)

2. Seventeen minus nine, times three, plus two, equals ? (26)

3. Three plus four, times six, minus three, equals ? (39)

4. Four times five, divided by 10 times seven, equals ? (14)

5. Sixty-six minus two, divided by eight, plus nine, equals ? (17)

6. Twenty-seven divided by nine, plus five, times seven, equals ? (56)

7. Thirty-one plus one, divided by eight, times six, equals ? (24)

8. Thirteen minus eight, times six, plus seven, equals ? (37)

9. Three times seven, plus three, divided by eight, equals ? (3)

10. Four times seven, plus eight, divided by four, equals ? (9)

Beginning with this lesson, there are also mental math problems on most of the systematic review pages in the *Epsilon Student Workbook.*

Subtracting Mixed Numbers
Regrouping and Unequal Denominators

In the examples, we separate the fractions, find a common denominator, rewrite the problem, and then perform the appropriate operation. If necessary, we regroup before subtracting.

Example 1

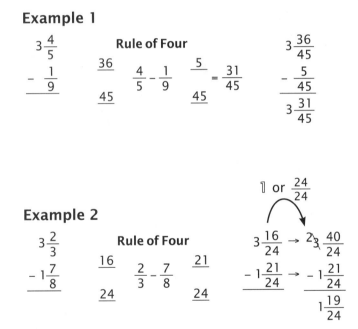

$$3\frac{4}{5}$$
$$-\ \frac{1}{9}$$

Rule of Four

$$\frac{36}{45} \qquad \frac{4}{5} - \frac{1}{9} \qquad \frac{5}{45} = \frac{31}{45}$$

$$3\frac{36}{45}$$
$$-\ \frac{5}{45}$$
$$3\frac{31}{45}$$

Example 2

$$3\frac{2}{3}$$
$$-1\frac{7}{8}$$

Rule of Four

$$\frac{16}{24} \qquad \frac{2}{3} - \frac{7}{8} \qquad \frac{21}{24}$$

$$1 \text{ or } \frac{24}{24}$$

$$3\frac{16}{24} \rightarrow 2\frac{40}{24}$$
$$-1\frac{21}{24} \rightarrow -1\frac{21}{24}$$
$$1\frac{19}{24}$$

There are two ways to solve Example 3. The first way uses regrouping, and the second way employs the "same difference theorem." Use the method that is more comfortable for you.

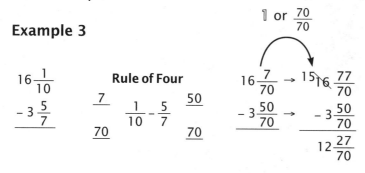

Example 3

$$16\frac{1}{10}$$
$$-3\frac{5}{7}$$

Rule of Four

$$\frac{7}{70} \quad \frac{1}{10} - \frac{5}{7} \quad \frac{50}{70}$$

1 or $\frac{70}{70}$

$$16\frac{7}{70} \to {}^{15}16\frac{77}{70}$$
$$-3\frac{50}{70} \to -3\frac{50}{70}$$
$$\rule{2cm}{0.4pt}$$
$$12\frac{27}{70}$$

The question is, "What do I have to add to 50/70 to make it a whole number?" The answer is 20/70. Adding this to both fractions produces $16^{27}/_{70}$ minus 4, which equals $12^{27}/_{70}$.

$$16\frac{7}{70} \quad +\frac{20}{70} \qquad 16\frac{27}{70}$$
$$-3\frac{50}{70} \quad +\frac{20}{70} \qquad -4$$
$$\rule{3cm}{0.4pt}$$
$$12\frac{27}{70}$$

Dividing Fractions and Mixed Numbers
Using the Reciprocal

Multiplying by the reciprocal of a number is the same as dividing by that number. A *reciprocal* is formed when the numerator and denominator switch places. The reciprocal of 2/3 is 3/2. The reciprocal of 4/11 is 11/4. Remember that any number may be written in fractional form as that number over one. For example, 2 may be written as 2/1, so the reciprocal of 2 is 1/2.

Using the reciprocal to divide is often referred to as "invert and multiply." I prefer to tell students that dividing by a fraction is equivalent to multiplying by the reciprocal of that fraction. Study the examples to see why this is true.

Example 1

$$\frac{2}{3} \times \frac{3}{2} = \frac{6}{6} = 1$$

$$\frac{4}{11} \times \frac{11}{4} = \frac{44}{44} = 1$$

The skill of finding the reciprocal will be useful in algebra because multiplying a number by its reciprocal always makes one.

Example 2

division $8 \div 2 = 4$

multiplication by the reciprocal $\frac{8}{1} \div \frac{2}{1} = \frac{8}{1} \times \frac{1}{2} = \frac{8}{2} = 4$

In the first line of Example 2, eight is divided by two to get four. In the second line, eight is multiplied by the reciprocal of two, which is one half. The result is four in both cases. Both methods produce the same answer.

Students have learned to divide by making the denominators the same and dividing the numerators. That method will always work; multiplying by the reciprocal just adds a second option. The following examples show division problems being solved using both methods.

Example 3

Solve: $\dfrac{4}{5} \div \dfrac{1}{3}$

Option 1: $\dfrac{4}{5} \div \dfrac{1}{3} = \dfrac{12}{15} \div \dfrac{5}{15} = \dfrac{12}{5} = 2\dfrac{2}{5}$

Option 2: $\dfrac{4}{5} \div \dfrac{1}{3} = \dfrac{4}{5} \times \dfrac{3}{1} = \dfrac{12}{5} = 2\dfrac{2}{5}$

There are advantages to both methods. After using the Rule of Four to make the denominators the same in Option 1, you may use the numerators to estimate the answer. Option 2 is a quick and simple way to solve a division problem with fractions.

When dividing mixed numbers, first convert them to improper fractions and then follow the same procedure.

Example 4

Solve: $1\dfrac{1}{3} \div \dfrac{2}{5}$

Option 1: $1\dfrac{1}{3} \div \dfrac{2}{5} = \dfrac{4}{3} \div \dfrac{2}{5} = \dfrac{20}{15} \div \dfrac{6}{15} = \dfrac{20}{6} = 3\dfrac{2}{6} = 3\dfrac{1}{3}$

Option 2: $1\dfrac{1}{3} \div \dfrac{2}{5} = \dfrac{4}{3} \div \dfrac{2}{5} = \dfrac{4}{3} \times \dfrac{5}{2} = \dfrac{20}{6} = 3\dfrac{2}{6} = 3\dfrac{1}{3}$

Solving for an Unknown
Using the Multiplicative Inverse; Word Problems

What can we multiply by 3/4 to make one? The answer is the reciprocal of 3/4, which is 4/3; $3/4 \times 4/3 = 1$. What can we multiply by five to make one? The answer is the reciprocal of 5/1, which is 1/5; $1/5 \times 5/1 = 1$. The *multiplicative inverse* is a number that, when multiplied by the given number, produces a product of one. The multiplicative inverse is the same as the reciprocal. The multiplicative inverse, or reciprocal, of 3/7 is 7/3.

This concept of making one is useful when we want to solve an equation with a number in front of the unknown, or variable. The number just before the letter that represents the unknown (for example, the 3 in 3X or the 5 in 5Y) is referred to as the *coefficient* of the variable. Remember that 3X means "three times X," and 5Y means "five times Y."

The primary objective to solving this kind of equation is to get the variable with a coefficient of one. Then we can write it without the coefficient as $1 \cdot X$, which is the same as X. If we multiply both sides of an equation by the same number, we do not change the value of the equation. Study the following examples.

Example 1

$$3X = 12$$
$$\frac{1}{3} \cdot 3X = \frac{12}{1} \cdot \frac{1}{3}$$
$$X = \frac{12}{3} = 4$$

1/3 times 3 equals 3/3, which is 1.

$$\frac{1}{3} \times 3 = \frac{1}{3} \times \frac{3}{1} = \frac{3}{3} = 1$$

Multiply both sides by 1/3, the reciprocal, or multiplicative inverse, of 3, to make a coefficient of 1. Then write the X without a coefficient and solve to find the value of X.

Example 2

$$5A = 35$$

$$\frac{1}{5} \cdot 5A = 35 \cdot \frac{1}{5}$$ 1/5 times 5 equals 5/5, which is 1.

$$A = \frac{35}{5} = 7 \qquad \frac{1}{5} \times 5 = \frac{1}{5} \times \frac{5}{1} = \frac{5}{5} = 1$$

Multiply both sides by 1/5, the multiplicative inverse of 5, to make a coefficient of 1. Then write the A without a coefficient and to find the value of A.

Always check your work by replacing the unknown with the solution. Remember our primary objective: to find the value of the unknown that will satisfy the equation. Here are checks for the previous examples. The solutions are placed in parentheses.

Example 1 check

$$3X = 12$$
$$3(4) = 12$$
$$12 = 12$$

Example 2 check

$$5A = 35$$
$$5(7) = 35$$
$$35 = 35$$

Word Problems

Here are a few more multi-step word problems to try. You may want to read and discuss these with your student as you work out the solutions together. Again, the purpose is to stretch, not to frustrate. If you do not think the student is ready, you may want to come back to these later.

The answers are at the end of the solutions section at the back of this book.

1. Mom mixed 2½ pounds of apples, 1⅛ pounds of grapes, and 1¼ pounds of pears for a salad. After setting aside 1½ pounds of salad for today, she divided the rest of the salad equally into three containers. What is the weight of the salad in one container?

2. Peter wants to put five fish in his aquarium. Three of the fish need one fourth of a cubic foot of water apiece, and two of them need one third of a cubic foot of water each. The dimensions of Peter's aquarium are 1 ft × 1 ft × 2 ft. Does he have room to add another fish that needs two thirds of a cubic foot of water? Assume that the aquarium is filled to the top.

3. Debbie had 5½ yards of ribbon. She cut it into pieces that were each 1½ yards long. How many inches long is the leftover piece? Note that the fraction that indicates the leftover piece is a fraction of a piece, not a fraction of the original amount.

Multiplying Mixed Numbers
Three Fractions; Canceling

When multiplying fractions, we can find the answer in two different ways. We have already learned that we can multiply the numerator and the denominator and then simplify the answer, if necessary, as in Example 1. Another method is a shortcut in which we simplify the fraction in the middle of the problem instead of at the end. This process is often called canceling. Example 2 shows the solution to the problem in Example 1, using this shortcut. Remember that multiplication is commutative, which means that the order of the factors may be changed without affecting the product.

Example 1

$$\frac{2}{3} \times \frac{5}{8} = \frac{10}{24} \div \frac{2}{2} = \frac{5}{12}$$

Example 2

$$\frac{2}{3} \times \frac{5}{8} = \frac{\overset{1}{2} \times 5}{3 \times \underset{4}{8}} = \frac{1 \times 5}{3 \times 4} = \frac{5}{12}$$

We can divide the 2 in the numerator by 2 to get 1, and we can divide the 8 in the denominator by 2 to get 4. Then we multiply across the numerator and the denominator: $1 \times 5 = 5$, and $3 \times 4 = 12$.

Both methods produce the same answer. Example 1 multiplies and then divides by a common factor of two to simplify. Example 2 divides by two in the middle of the problem and then multiplies to find the solution.

We can either divide by a common factor in the middle of the expression (canceling) or at the end of the expression (simplifying). Another way to think of this is to rearrange the common factors. Look at Example 3. Now we have a three above a three and five above a five. Since a number divided by itself is one, these pairs of numbers will both equal one.

Example 3

$$\frac{2}{3} \times \frac{5}{8} \times \frac{3}{5} = \frac{2 \times 5 \times 3}{3 \times 8 \times 5} = \frac{2 \times 3 \times 5}{8 \times 3 \times 5} = \frac{\cancel{2}^1 \times \cancel{3}^1 \times \cancel{5}^1}{\cancel{8}_4 \times \cancel{3}_1 \times \cancel{5}_1} = \frac{1}{4}$$

Now we can solve the problem without rearranging the factors. Simply divide the numerator and the denominator by a common factor. Example 4 shows the same problem from Example 3 worked out step by step in this manner. We can do all the canceling without rewriting each step.

Example 4

$$\frac{2}{3} \times \frac{5}{8} \times \frac{3}{5} = \frac{2}{\cancel{3}_1} \times \frac{5}{8} \times \frac{\cancel{3}^1}{5} = \frac{2}{1} \times \frac{\cancel{5}^1}{8} \times \frac{1}{\cancel{5}_1}$$

$$= \frac{\cancel{2}^1}{1} \times \frac{1}{\cancel{8}_4} \times \frac{1}{1} = \frac{1}{1} \times \frac{1}{4} \times \frac{1}{1} = \frac{1}{4}$$

Remember our first option, which is to multiply all the factors in the numerator and the denominator and then to simplify the final answer, as in Example 5 below.

Example 5

$$\frac{2}{3} \times \frac{5}{8} \times \frac{3}{5} = \frac{30}{120} \div \frac{30}{30} = \frac{1}{4}$$

Next we will look at problems involving mixed numbers. Here our first step is to convert the mixed numbers to improper fractions. In Example 6, the mixed number 1⅓ becomes 4/3, and 2½ becomes 5/2.

Example 6

$$1\frac{1}{3} \times 2\frac{1}{2} \times \frac{3}{5} = \frac{4}{\cancel{3}_1} \times \frac{5}{2} \times \frac{\cancel{3}^1}{5}$$

$$= \frac{4}{1} \times \frac{\cancel{5}^1}{2} \times \frac{1}{\cancel{5}_1} = \frac{\cancel{4}^2}{1} \times \frac{1}{\cancel{2}_1} \times \frac{1}{1} = \frac{2}{1} = 2$$

There may be more than one way to solve the problem.

$$1\frac{1}{3} \times 2\frac{1}{2} \times \frac{3}{5} = \frac{\cancel{4}^{2}}{\cancel{3}_{1}} \times \frac{\cancel{5}^{1}}{\cancel{2}_{1}} \times \frac{\cancel{3}^{1}}{\cancel{5}_{1}} = \frac{2}{1} = 2$$

or $1\frac{1}{3} \times 2\frac{1}{2} \times \frac{3}{5} = \frac{4}{3} \times \frac{5}{2} \times \frac{3}{5} = \frac{60}{30} \div \frac{30}{30} = \frac{2}{1} = 2$

Example 7

$$1\frac{1}{5} \times 3\frac{1}{3} \times 1\frac{3}{4} = \frac{6}{5} \times \frac{\cancel{10}^{5}}{3} \times \frac{7}{\cancel{4}_{2}} = \frac{6}{\cancel{5}_{1}} \times \frac{\cancel{5}^{1}}{3} \times \frac{7}{2}$$

$$= \frac{\cancel{6}^{2}}{1} \times \frac{1}{\cancel{3}_{1}} \times \frac{7}{2} = \frac{\cancel{2}^{1}}{1} \times \frac{1}{1} \times \frac{7}{\cancel{2}_{1}} = 7$$

There may be more than one way to solve the problem.

$$1\frac{1}{5} \times 3\frac{1}{3} \times 1\frac{3}{4} = \frac{\cancel{6}^{2}}{\cancel{5}_{1}} \times \frac{\cancel{10}^{1}}{\cancel{3}_{1}} \times \frac{7}{\cancel{4}_{\cancel{2}_{1}}} = 7$$

or $1\frac{1}{5} \times 3\frac{1}{3} \times 1\frac{3}{4} = \frac{6}{5} \times \frac{10}{3} \times \frac{7}{4} = \frac{420}{60} \div \frac{60}{60} = \frac{7}{1} = 7$

Notice that you can keep on canceling as long as you have numbers in the numerator and the denominator that share the same factor.

In the third part of Example 7, the six is divided by three, which leaves two in the numerator. Then the four in the denominator and 10 in the numerator are both divided by two, which leaves two in the denominator and five in the numerator. We still have a two in the numerator and a two in the denominator. We can divide again by two. Even though this process is called canceling, we are not simply throwing out numbers. Instead, we are simplifying the fractions to other fractions with the same value.

Area and Circumference of a Circle
Using Pi ≈ 22/7

The formula for the area of a circle is πr^2. The name of the symbol π is spelled *pi* and is pronounced in English like "pie." It represents a specific value that is a little more than three. The r represents the **radius,** which is the distance from the center of the circle to the edge of the circle.

The number π is not a whole number, since it comes between 3 and 4. For a long time, people tried to find a fraction between 3 and 4 that is exactly equal to π, but it turns out that there is none. In this book, we will use the fraction 22/7, which is a good approximation. (You may see the decimal value of π given as 3.14, which is also a rounded value. We will discuss this more when we have studied decimals.) To help you understand and remember the value of π, study Figure 1 below. Remember that area is always given in square units.

Figure 1
Imagine a square with a side of length r and an area of r × r, or r^2. The area of the circle is a little more than the area of three of the squares.

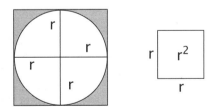

When a number has a small 2 above it, as in r^2, it indicates that you use the number as a factor two times. The small number is called an exponent. We will discuss exponents in more detail in *Zeta*.

$$3^2 = 3 \times 3, = 9$$
$$5^2 = 5 \times 5, = 25$$

Example 1
Find the area of the circle.

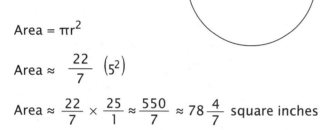

$$\text{Area} = \pi r^2$$

$$\text{Area} \approx \frac{22}{7}\left(5^2\right)$$

$$\text{Area} \approx \frac{22}{7} \times \frac{25}{1} \approx \frac{550}{7} \approx 78\frac{4}{7} \text{ square inches}$$

Example 2
Find the area of the circle.

$$\text{Area} = \pi r^2$$

$$\text{Area} \approx \frac{22}{7}\left(7^2\right)$$

$$\text{Area} \approx \frac{22}{7} \times \frac{49}{1} \approx \frac{22}{\underset{1}{\cancel{7}}} \times \frac{\overset{7}{\cancel{49}}}{1} \approx \frac{154}{1} \approx 154 \text{ square feet}$$

Circumference of a Circle

Rectangles have area and perimeter. Circles have area and circumference. The *circumference* is the distance around the outside of a circle and corresponds to the perimeter of a rectangle. The value of π times the diameter of a circle equals the circumference of that circle.

Since the diameter is twice the radius, another formula for the circumference of a circle is 2πr. Circumference is always given with units of length, such as inches or centimeters.

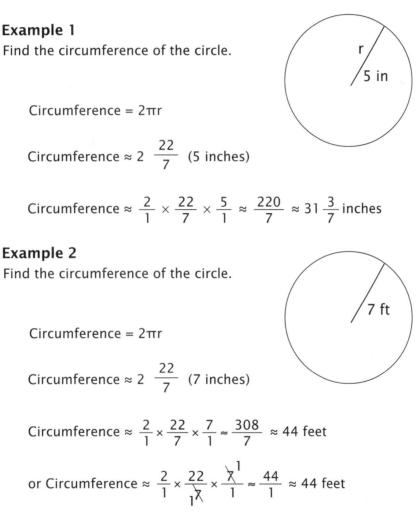

Example 1
Find the circumference of the circle.

Circumference = 2πr

Circumference ≈ 2 $\dfrac{22}{7}$ (5 inches)

Circumference ≈ $\dfrac{2}{1} \times \dfrac{22}{7} \times \dfrac{5}{1} \approx \dfrac{220}{7} \approx 31\dfrac{3}{7}$ inches

Example 2
Find the circumference of the circle.

Circumference = 2πr

Circumference ≈ 2 $\dfrac{22}{7}$ (7 inches)

Circumference ≈ $\dfrac{2}{1} \times \dfrac{22}{7} \times \dfrac{7}{1} \approx \dfrac{308}{7} \approx 44$ feet

or Circumference ≈ $\dfrac{2}{1} \times \dfrac{22}{1\cancel{7}} \times \dfrac{\cancel{7}^{1}}{1} \approx \dfrac{44}{1} \approx 44$ feet

Solving for an Unknown
Fractional Coefficients

This lesson is similar to lesson 26, but now the coefficients are fractions rather than whole numbers. After solving for the unknown, always check the answer. Study the examples.

Example 1

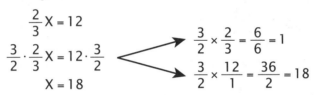

$$\frac{2}{3}X = 12$$

$$\frac{3}{2} \cdot \frac{2}{3}X = 12 \cdot \frac{3}{2}$$

$$X = 18$$

$$\frac{3}{2} \times \frac{2}{3} = \frac{6}{6} = 1$$

$$\frac{3}{2} \times \frac{12}{1} = \frac{36}{2} = 18$$

Multiply both sides by 3/2, the reciprocal of 2/3, to get X by itself.

Example 1 check

$$\frac{2}{3}X = 12$$

$$\frac{2}{3}(18) = 12 \longrightarrow \frac{2}{3} \times \frac{18}{1} = \frac{36}{3} = 12$$

$$12 = 12$$

Always check your work by replacing the unknown with the solution. Remember our primary objective: to find the value of the unknown that will satisfy the equation.

Example 2

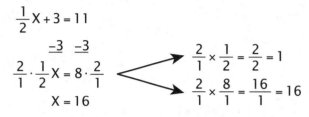

$$\frac{1}{2}X + 3 = 11$$

$$\underline{-3 \quad -3}$$

$$\frac{2}{1} \cdot \frac{1}{2}X = 8 \cdot \frac{2}{1}$$

$$X = 16$$

$$\frac{2}{1} \times \frac{1}{2} = \frac{2}{2} = 1$$

$$\frac{2}{1} \times \frac{8}{1} = \frac{16}{1} = 16$$

Multiply both sides by 2/1, the reciprocal of 1/2, to get X by itself.

Example 2 check

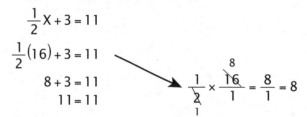

$$\frac{1}{2}X + 3 = 11$$

$$\frac{1}{2}(16) + 3 = 11$$

$$8 + 3 = 11$$

$$11 = 11$$

$$\frac{1}{\cancel{2}_{1}} \times \frac{\cancel{16}^{8}}{1} = \frac{8}{1} = 8$$

Fractions to Decimals to Percents

A *decimal* is a way of writing a fraction on one line (without a numerator or denominator). In our decimal system, there are values assigned to each place or position. This concept is called *place value* (see Figure 1).

Figure 1

$$\overline{}\ \overline{}\ \overline{}\ \overline{}\ \cdot\ \overline{\phantom{\frac{1}{10}}}\ \overline{\phantom{\frac{1}{100}}}\ \overline{\phantom{\frac{1}{1,000}}}$$

$$1,000 \quad 100 \quad 10 \quad 1 \qquad \frac{1}{10} \quad \frac{1}{100} \quad \frac{1}{1,000}$$

All numbers in the decimal system have at least one digit (0 through 9), each with its own place value. The number 142 may be written in *expanded notation* as 1×100 + 4×10 + 2×1. The digits are 1, 4, and 2, and the values are hundreds, tens, and units. In Figure 2 below, 142 is written in decimal notation.

Figure 2

$$\overline{}\ \overline{1}\ \overline{4}\ \overline{2}\ \cdot\ \overline{\phantom{\frac{1}{10}}}\ \overline{\phantom{\frac{1}{100}}}\ \overline{\phantom{\frac{1}{1,000}}}$$

$$1,000 \quad 100 \quad 10 \quad 1 \qquad \frac{1}{10} \quad \frac{1}{100} \quad \frac{1}{1,000}$$

The number 0.5 is a decimal number. Some call it a decimal fraction. It is read "five tenths" and is written in expanded notation as 5 × 1/10. Figure 3 shows its position in decimal notation. (The zero before the decimal reinforces that there are no units and that the number is less than 1.)

Figure 3

$$\overline{}\ \overline{}\ \overline{}\ \overline{0}\ \cdot\ \overline{5}\ \overline{\phantom{\frac{1}{100}}}\ \overline{\phantom{\frac{1}{1,000}}}$$

$$1,000 \quad 100 \quad 10 \quad 1 \qquad \frac{1}{10} \quad \frac{1}{100} \quad \frac{1}{1,000}$$

A fraction may also be written using expanded notation. Two fifths (2/5) is the same as 2 × 1/5. We can't write 2/5 as a decimal because there is no fifths place in the decimal system. We must change the denominator to 10, 100, 1,000, or some other multiple of ten. To show how this works, build 2/5 and then place the 1/10 overlay on top, as shown in Example 1.

Example 1
Change 2/5 to a decimal.

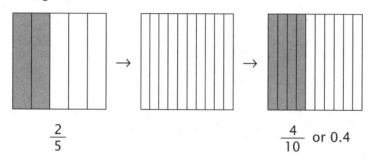

$$\frac{2}{5}$$ $$\frac{4}{10} \text{ or } 0.4$$

We have just renamed 2/5 as 4/10. The number is written as 0.4 in decimal notation. The zero in the units place shows that the number is less than 1. The number is read as "four tenths."

Example 2
Change 1/2 to a decimal.

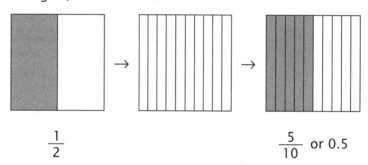

$$\frac{1}{2}$$ $$\frac{5}{10} \text{ or } 0.5$$

The key is to change the denominator to a power of 10 so that the fraction can be written as a decimal.

Example 3

Change 3/4 to a decimal.

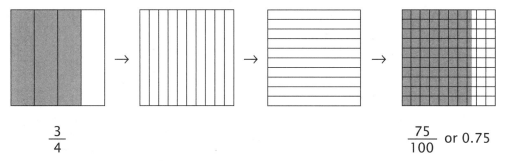

$$\frac{3}{4} \qquad \frac{75}{100} \text{ or } 0.75$$

To illustrate this clearly, make 3/4 with the fraction kit and then take off the clear fourths overlay. Place both tenths overlays on top of 3/4, with one overlay turned sideways to make a 100 grid. The fraction 3/4 cannot be renamed as tenths, so we changed it to hundredths instead. We have seven rows of 10 that equal 70 and then a half row of 10, which equals five. Since 70 + 5 = 75, the fraction 3/4 is the same as the fraction 75/100 and the decimal 0.75.

The fractions listed below may be used to illustrate changing a fraction to a decimal. They work out exactly. Other fractions, such as the thirds and the sixths, don't come out evenly, but the overlays will give a good approximation.

$$\frac{1}{5}, \frac{2}{5}, \frac{3}{5}, \frac{4}{5}, \frac{5}{5}, \frac{1}{2}, \frac{1}{4}, \frac{3}{4}$$

To change a fraction or a decimal to a percent, we must recognize that *percent* means "per hundred." First we transform a fraction or decimal to hundredths (1/100) as in Example 3. Watch the progression of the symbols in Example 4.

Example 4

Change 3/4 to 0.75 to a percent.

$$\frac{3}{4} = 0.75 = \frac{75}{100} \rightarrow \frac{75}{00}\,^{/} \rightarrow \frac{75\,°/}{0} \rightarrow \frac{75\%}{} \rightarrow 75\%$$

We took the one and the two zeros from the denominator of the fraction and used them to form the percent symbol. Percents are simply a different way of writing hundredths. Once a fraction or decimal is changed to hundredths, changing to a percent is quick and easy.

Comparing Decimals

It is easier to compare decimals if they have the same number of places. In other words, you can easily compare tenths to tenths, hundredths to hundredths, and thousandths to thousandths.

Example 5
Which decimal is greater: 0.5 or 0.6?

Both decimals are expressed as tenths. Since 6 is greater than 5, the answer is 0.6.

Using an inequality sign, we can write the answer as 0.6 > 0.5.

Example 6
Which decimal is less: 3.25 or 3.45?

The number in the units place is the same for both numbers, so we look at the decimal part of the number. Both decimals are expressed as hundredths. Since 25 is less than 45, the answer is 3.25.

Using an inequality sign, we can write the answer as 3.25 < 3.45.

Example 7
Compare the decimals: 0.8 and 0.80

It is easier to compare decimals if they have the same number of places. We can think of these decimals as 8/10 and 80/100. We know from our study of fractions that 80/100 may be simplified to 8/10.

8/10 = 8/100, so 0.8 = 0.80. The decimals have the same value.

In Example 7, notice that 0.8 could also be written as 0.80 without changing its value. If the student understands this, it is not necessary to change the decimals to fractions in order to compare them.

More Solving for an Unknown

In lesson 28 we used multiplication, addition, and subtraction to solve for an unknown. We concentrated on problems where the coefficient of the unknown was a fraction. In this lesson the number being added to (or subtracted from) both sides of the equation is also a fraction.

Remember that, when we multiply by the reciprocal of 2/3, we are using a multiplicative inverse.

Example 1

$$\frac{3}{5}G - \frac{1}{2} = 5\frac{1}{2}$$
$$+\frac{1}{2} \quad +\frac{1}{2}$$
$$\frac{3}{5}G + 0 = 6$$
$$\frac{5}{3} \cdot \frac{3}{5}G = 6 \cdot \frac{5}{3}$$
$$G = \frac{30}{3}$$
$$G = 10$$

Add the opposite of (–1/2), which is (+1/2), to both sides.

Multiply both sides by 5/3, which is the reciprocal of 3/5.

Example 1 check

$$\frac{3}{5}\bigcirc - \frac{1}{2} = 5\frac{1}{2}$$

$$\frac{3}{5}(10) - \frac{1}{2} = 5\frac{1}{2}$$

$$\frac{30}{5} - \frac{1}{2} = 5\frac{1}{2}$$

$$6 - \frac{1}{2} = 5\frac{1}{2}$$

$$5\frac{1}{2} = 5\frac{1}{2}$$

Example 2

$$\frac{1}{4}M + \frac{5}{8} = \frac{3}{4}$$

$$-\frac{5}{8} \quad -\frac{5}{8}$$

$$\frac{1}{4}M + 0 = \frac{3}{4} - \frac{5}{8}$$

$$\frac{1}{4}M = \frac{24}{32} - \frac{20}{32}$$

$$\frac{4}{1} \cdot \frac{1}{4}M = \frac{4}{32} \cdot \frac{4}{1}$$

$$M = \frac{16}{32}$$

$$M = \frac{1}{2}$$

Subtract 5/8 from both sides.

Multiply both sides by 4/1, which is the reciprocal of 1/4.

Example 2 check

$$\frac{1}{4} \cdot \frac{1}{2} + \frac{5}{8} = \frac{3}{4}$$

$$\frac{1}{8} + \frac{5}{8} = \frac{3}{4}$$

$$\frac{6}{8} = \frac{3}{4}$$

$$\frac{3}{4} = \frac{3}{4}$$

Finding the Area of a Trapezoid

Finding the area of a trapezoid is taught in *Delta* and again in *Geometry*. This lesson is for those who started Math-U-See after the *Delta* level. It may also be used as a review.

A *quadrilateral* is a closed two-dimensional shape with exactly four straight sides. A *trapezoid* is a quadrilateral that has exactly one pair of sides that are parallel to each other. In the picture below, notice that the top and bottom sides are parallel but the other two sides are not. The top and bottom sides are called the *bases.*

Finding the area of a trapezoid is similar to finding the area of a rectangle. To find the area of a rectangle, multiply the base by the height. To find the area of a trapezoid, multiply the average base by the height. Look at Figure 1 to see where this formula originates.

Figure 1

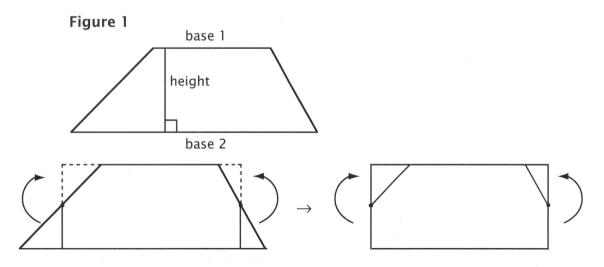

On the left and right sides in Figure 1, we chose the midpoint of the sides and drew a line perpendicular to the bases through it. This made one small right triangle on each side. Then we pivoted the lower triangles on the midpoint and swung them up to make a rectangle out of the trapezoid. The resulting base is the average of the top and bottom bases and is shown by connecting the midpoints on the two sides.

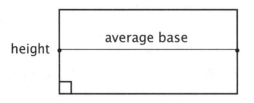

The average base is found by adding the top and bottom bases and dividing by two. Then this result is multiplied by the height to find the area.

The formula for this process is $\frac{b_1 + b_2}{2} \times h$.

Example 1
Find the area of the trapezoid.

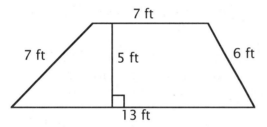

The average base is (7 ft + 13 ft) ÷ 2 = 10 ft. The height is 5 ft.
The area is 10 ft × 5 ft = 50 square feet, or 50 ft^2.

Example 2
Find the area of the trapezoid.

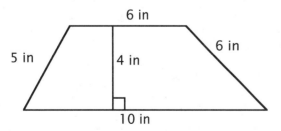

The average base is (6 in +10 in) ÷ 2 = 8 inches. The height is 4 inches.
The area is 8 inches × 4 inches = 32 square inches, or 32 in^2.

Student Solutions

Lesson Practice 1A

1. done

2. $6; 2; 1; \frac{1}{2}$ of 6 is 3.

3. $10; 5; 4; \frac{4}{5}$ of 10 is 8.

4. $9; 3; 2; \frac{2}{3}$ of 9 is 6.

5. $8; 4; 3; \frac{3}{4}$ of 8 is 6.

Lesson Practice 1B

1. $\frac{3}{5}$ of 15 is 9.

2. $\frac{2}{3}$ of 18 is 12.

3. $\frac{5}{8}$ of 8 is 5.

4. $20 \div 5 = 4$
   $4 \times 3 = 12$

5. $6 \div 3 = 2$
   $2 \times 2 = 4$

6. $8 \div 2 = 4$
   $4 \times 1 = 4$

7. $4 \div 4 = 1$
   $1 \times 3 = 3$

8. $6 \div 3 = 2$
   $2 \times 1 = 2$

9. $8 \div 4 = 2$
   $2 \times 1 = 2$

10. $10 \div 5 = 2$
    $2 \times 4 = 8$

11. $6 \div 2 = 3$
    $3 \times 1 = 3$

12. $12 \div 3 = 4$
    $4 \times 2 = 8$

13. $10 \div 5 = 2$
    $2 \times 2 = 4$

14. $12 \div 4 = 3$
    $3 \times 1 = 3$ eggs

15. $10 \div 2 = 5$
    $5 \times 1 = 5$ boys

Lesson Practice 1C

1. $\frac{1}{3}$ of 12 is 4.

2. $\frac{3}{4}$ of 20 is 15.

3. $\frac{7}{9}$ of 18 is 14.

4. $16 \div 4 = 4$
   $4 \times 2 = 8$

5. $10 \div 5 = 2$
   $2 \times 1 = 2$

6. $18 \div 6 = 3$
   $3 \times 5 = 15$

7. $9 \div 3 = 3$
   $3 \times 1 = 3$

8. $12 \div 4 = 3$
   $3 \times 3 = 9$

9. $10 \div 5 = 2$
   $2 \times 3 = 6$

10. $4 \div 2 = 2$
    $2 \times 1 = 2$

11. $12 \div 3 = 4$
    $4 \times 1 = 4$

12. $12 \div 2 = 6$
    $6 \times 1 = 6$

13. $16 \div 8 = 2$
    $2 \times 7 = 14$

14. $14 \div 7 = 2$
    $2 \times 3 = 6$ days

15. $20 \div 5 = 4$
    $4 \times 4 = 16$ tulip plants

Systematic Review 1D

1. $\frac{2}{5}$ of 15 is 6.

2. $\frac{1}{3}$ of 9 is 3.

3. $18 \div 9 = 2$
   $2 \times 5 = 10$

4. $10 \div 5 = 2$
   $2 \times 2 = 4$

5. $8 \div 4 = 2$
   $2 \times 3 = 6$

6. $30 \div 6 = 5$
   $5 \times 5 = 25$
7. $48 \div 8 = 6$
   $6 \times 7 = 42$
8. $24 \div 6 = 4$
   $4 \times 3 = 12$
9. $35 \div 7 = 5$
   $5 \times 4 = 20$
10. $10 \div 5 = 2$
    $2 \times 3 = 6$
11. done
12. $20 + 10 + 20 + 10 = 60$ ft
13. $15 + 7 + 15 + 7 = 44$ in
14. $60 \div 3 = 20$
    $20 \times 2 = 40$ minutes
15. $25 \div 5 = 5$
    $5 \times 3 = 15$ girls

Systematic Review 1E

1. $\frac{4}{5}$ of 20 is 16.
2. $\frac{1}{2}$ of 6 is 3.
3. $10 \div 5 = 2$
   $2 \times 5 = 10$
4. $20 \div 4 = 5$
   $5 \times 3 = 15$
5. $24 \div 6 = 4$
   $4 \times 5 = 20$
6. $56 \div 7 = 8$
   $8 \times 1 = 8$
7. $12 \div 3 = 4$
   $4 \times 2 = 8$
8. $10 \div 5 = 2$
   $2 \times 2 = 4$
9. $8 \div 4 = 2$
   $2 \times 1 = 2$
10. $32 \div 8 = 4$
    $4 \times 3 = 12$
11. $25 + 13 + 25 + 13 = 76$ in
12. $32 + 16 + 32 + 16 = 96$ ft
13. $78 + 39 + 78 + 39 = 234$ in

14. $\$16 \div 2 = \$8$
    $\$8 \times 1 = \$8$
15. $20 + 30 + 20 + 30 = 100$ ft
16. $56 \div 7 = 8$
    $8 \times 4 = 32$ plates
17. $6 + 11 + 6 + 11 = 34$ ft
18. $18 \div 9 = 2$
    $2 \times 7 = 14$ students

Systematic Review 1F

1. $\frac{3}{6}$ of 24 = 12.
2. $\frac{2}{4}$ of 8 = 4.
3. $21 \div 3 = 7$
   $7 \times 2 = 14$
4. $6 \div 3 = 2$
   $2 \times 1 = 2$
5. $12 \div 3 = 4$
   $4 \times 1 = 4$
6. $12 \div 4 = 3$
   $3 \times 3 = 9$
7. $25 \div 5 = 5$
   $5 \times 1 = 5$
8. $42 \div 6 = 7$
   $7 \times 5 = 35$
9. $54 \div 9 = 6$
   $6 \times 1 = 6$
10. $16 \div 8 = 2$
    $2 \times 5 = 10$
11. $9 + 6 + 9 + 6 = 30$ in
12. $18 + 10 + 18 + 10 = 56$ ft
13. $43 + 27 + 43 + 27 = 140$ in
14. $36 \div 9 = 4$
    $4 \times 5 = 20$ candies
15. $36 \div 9 = 4$
    $4 \times 4 = 16$
    $16 + 20 = 36$ candies
16. $\$12.00 + \$20.00 + \$8.50 = \$40.50$
17. $\$40.50 - \$14.50 = \$26.00$
18. $12 \div 4 = 3$
    $3 \times 3 = 9$ months

Lesson Practice 2A

1. done
2. $\frac{3}{5}$; three fifths
3. $\frac{1}{5}$; one fifth
4. $\frac{2}{6}$; two sixths
5. $\frac{2}{3}$; two thirds
6. $\frac{2}{2}$; two halves
7. $\frac{3}{4}$; three fourths
8. $\frac{4}{6}$; four sixths
9. done
10. two fourths; 4 sections, 2 shaded

11. one third; 3 sections, 1 shaded

12. two halves; 2 sections, both shaded

13. done
14. $\frac{2}{3}$; 3 sections, 2 shaded

15. $\frac{1}{4}$; 4 sections, 1 shaded

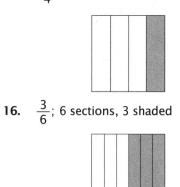

16. $\frac{3}{6}$; 6 sections, 3 shaded

Lesson Practice 2B

1. $\frac{3}{3}$; three thirds
2. $\frac{1}{6}$; one sixth
3. $\frac{2}{5}$; two fifths
4. $\frac{6}{6}$; six sixths
5. $\frac{1}{3}$; one third
6. $\frac{1}{2}$; one half
7. $\frac{2}{4}$; two fourths
8. $\frac{4}{5}$; four fifths
9. three thirds; 3 sections, all shaded

10. one sixth; 6 sections, 1 shaded

11. one half; 2 sections, 1 shaded

12. four sixths; 6 sections, 4 shaded

13. $\frac{3}{5}$; 5 sections, 3 shaded

14. $\frac{4}{4}$; 4 sections, all shaded

15. $\frac{4}{5}$; 5 sections, 4 shaded

16. $\frac{5}{6}$; 6 sections, 5 shaded

Lesson Practice 2C

1. $\frac{2}{3}$; two thirds

2. $\frac{5}{6}$; five sixths

3. $\frac{3}{5}$; three fifths

4. $\frac{5}{6}$; five sixths

5. $\frac{2}{2}$; two halves

6. $\frac{1}{3}$; one third

7. $\frac{3}{4}$; three fourths

8. $\frac{1}{5}$; one fifth

9. two sixths; 6 sections, 2 shaded

10. one fourth; 4 sections, 1 shaded

11. two thirds; 3 sections, 2 shaded

12. four fifths; 5 sections, 4 shaded

13. $\frac{1}{5}$; 5 sections, 1 shaded

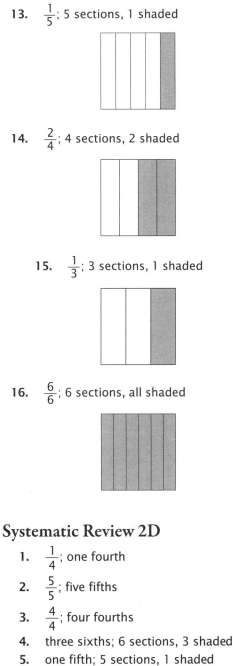

14. $\frac{2}{4}$; 4 sections, 2 shaded

15. $\frac{1}{3}$; 3 sections, 1 shaded

16. $\frac{6}{6}$; 6 sections, all shaded

Systematic Review 2D

1. $\frac{1}{4}$; one fourth

2. $\frac{5}{5}$; five fifths

3. $\frac{4}{4}$; four fourths

4. three sixths; 6 sections, 3 shaded

5. one fifth; 5 sections, 1 shaded

6. $\frac{5}{6}$; 6 sections, 5 shaded

7. $24 \div 3 = 8$; $8 \times 2 = 16$

8. $40 \div 4 = 10$; $3 \times 10 = 30$

9. $34 \div 2 = 17$; $1 \times 17 = 17$

10. done

11. $15 + 15 + 15 + 15 = 60$ ft

12. rectangle:
$12 + 24 + 12 + 24 = 72$ in
square:
$10 + 10 + 10 + 10 = 40$ in
combined:
$72 + 40 = 112$ in

13. $18 \div 9 = 2$
$2 \times 5 = 10$ players

14. $12 \div 4 = 3$
$3 \times 3 = 9$ bushes

Systematic Review 2E

1. $\frac{3}{6}$; three sixths

2. $\frac{5}{6}$; five sixths

3. $\frac{2}{5}$; two fifths

4. $\frac{1}{6}$; 6 sections, 1 shaded

5. $\frac{2}{3}$; 3 sections, 2 shaded

6. six sixths; 6 sections, all shaded

7. $36 \div 6 = 6$
$1 \times 6 = 6$

8. $72 \div 9 = 8$
$8 \times 4 = 32$

9. $39 \div 3 = 13$
$13 \times 2 = 26$

10. $4 + 17 + 4 + 17 = 42$ yd

11. $34 + 34 + 34 + 34 = 136$ ft

12. $18 + 18 + 18 + 18 = 72$ yd to start
$72 - 10 = 62$ yd left

13. $12 \div 4 = 3$
$3 \times 1 = 3$ months

14. $12 \div 3 = 4$
$4 \times 2 = 8$ months

15. $35 \div 7 = 5$
$5 \times 5 = 25$ camels

Systematic Review 2F

1. $\frac{2}{4}$; two fourths

2. $\frac{4}{5}$; four fifths

3. $\frac{2}{6}$; two sixths

4. $\frac{3}{5}$; 5 sections, 3 shaded

5. $\frac{2}{6}$; six sections, 2 shaded

6. three fourths; 4 sections, 3 shaded

7. $80 \div 2 = 40$; $1 \times 40 = 40$

8. $55 \div 5 = 11$; $11 \times 3 = 33$

9. $56 \div 8 = 7$; $7 \times 3 = 21$

10. $46 + 21 + 46 + 21 = 134$ yd

11. $13 + 13 + 13 + 13 = 52$ ft

12. $12 \div 6 = 2$
    $2 \times 1 = 2$ eggs

13. $20 \div 5 = 4$
    $4 \times 3 = 12$ people

14. $20 \div 5 = 4$
    $4 \times 2 = 8$ people

15. $\$10.50 + \$7.25 = \$17.75$

Lesson Practice 3A

1. $\frac{5}{5}$; five fifths

2. $\frac{1}{3}$; one third

3. $\frac{1}{4} + \frac{2}{4} = \frac{3}{4}$ or three fourths

4. $\frac{3}{6} - \frac{2}{6} = \frac{1}{6}$ or one sixth

5. $\frac{1}{6} + \frac{3}{6} = \frac{4}{6}$

6. $\frac{1}{5} + \frac{2}{5} = \frac{3}{5}$

7. $\frac{4}{6} + \frac{1}{6} = \frac{5}{6}$

8. $\frac{4}{6} - \frac{1}{6} = \frac{3}{6}$

9. $\frac{3}{4} - \frac{2}{4} = \frac{1}{4}$

10. $\frac{3}{6} - \frac{2}{6} = \frac{1}{6}$

11. $\frac{1}{4} + \frac{2}{4} = \frac{3}{4}$ of the letters

12. $\frac{1}{5} + \frac{3}{5} = \frac{4}{5}$ of the book

13. $\frac{6}{6} - \frac{2}{6} = \frac{4}{6}$ of the job left

14. $\frac{4}{5} - \frac{1}{5} = \frac{3}{5}$ of the students

Lesson Practice 3B

1. $\frac{4}{6}$; four sixths

2. $\frac{3}{5}$; three fifths

3. $\frac{2}{2}$; two halves

4. $\frac{2}{4}$; two fourths

5. $\frac{2}{4} + \frac{2}{4} = \frac{4}{4}$

6. $\frac{3}{6} + \frac{2}{6} = \frac{5}{6}$

7. $\frac{1}{5} + \frac{3}{5} = \frac{4}{5}$

8. $\frac{5}{6} - \frac{1}{6} = \frac{4}{6}$

9. $\frac{3}{4} - \frac{1}{4} = \frac{2}{4}$

10. $\frac{2}{3} - \frac{1}{3} = \frac{1}{3}$

11. $\frac{1}{4} + \frac{1}{4} = \frac{2}{4}$ of the laps

12. $\frac{10}{10} - \frac{1}{10} = \frac{9}{10}$ of her pay

13. $\frac{2}{5} + \frac{2}{5} = \frac{4}{5}$ of the trip completed

    $\frac{5}{5} - \frac{4}{5} = \frac{1}{5}$ of the trip left

14. $\frac{5}{6} - \frac{2}{6} = \frac{3}{6}$ of the job

Lesson Practice 3C

1. $\frac{1}{3} + \frac{2}{3} = \frac{3}{3}$ three thirds

2. $\frac{6}{6} - \frac{3}{6} = \frac{3}{6}$ three sixths

3. $\frac{4}{5}$; four fifths

4. $\frac{1}{4}$; one fourth

5. done

6. $\dfrac{2}{5} + \dfrac{3}{5} = \dfrac{5}{5}$

7. $\dfrac{6}{7} + \dfrac{1}{7} = \dfrac{7}{7}$

8. $\dfrac{2}{6} - \dfrac{1}{6} = \dfrac{1}{6}$

9. $\dfrac{5}{8} - \dfrac{4}{8} = \dfrac{1}{8}$

10. $\dfrac{6}{10} - \dfrac{1}{10} = \dfrac{5}{10}$

11. $\dfrac{3}{8} + \dfrac{1}{8} = \dfrac{4}{8}$ tsp of salt

12. $\dfrac{3}{12} + \dfrac{2}{12} = \dfrac{5}{12}$ of the pizza eaten

13. $\dfrac{12}{12} - \dfrac{5}{12} = \dfrac{7}{12}$ of the pizza left

14. $\dfrac{9}{10} - \dfrac{5}{10} = \dfrac{4}{10}$ of a load

Systematic Review 3D

1. $\dfrac{1}{4} + \dfrac{1}{4} = \dfrac{2}{4}$ two fourths

2. $\dfrac{5}{6} - \dfrac{1}{6} = \dfrac{4}{6}$ four sixths

3. $\dfrac{1}{2} + \dfrac{1}{2} = \dfrac{2}{2}$

4. $\dfrac{3}{10} + \dfrac{6}{10} = \dfrac{9}{10}$

5. $\dfrac{1}{6} + \dfrac{3}{6} = \dfrac{4}{6}$

6. $\dfrac{4}{6} - \dfrac{1}{6} = \dfrac{3}{6}$

7. $\dfrac{3}{4} - \dfrac{2}{4} = \dfrac{1}{4}$

8. $\dfrac{5}{9} - \dfrac{3}{9} = \dfrac{2}{9}$

9. $10 \div 5 = 2$
$2 \times 4 = 8$

10. $40 \div 8 = 5$
$5 \times 5 = 25$

11. $49 \div 7 = 7$
$7 \times 2 = 14$

12. done

13. $3 + 4 + 5 = 12$ ft

14. $54 + 39 + 78 = 171$ ft

15. $\dfrac{4}{9} + \dfrac{1}{9} = \dfrac{5}{9}$ of a box

16. $\dfrac{3}{8} + \dfrac{2}{8} = \dfrac{5}{8}$ of the problems

$\dfrac{8}{8} - \dfrac{5}{8} = \dfrac{3}{8}$ of the problems

17. $16 \div 2 = 8$
$8 \times 1 = 8$ people

18. $24 \div 6 = 4$
$4 \times 5 = 20$ pa□es

Systematic Review 3E

1. $\dfrac{3}{5}$; three fifths

2. $\dfrac{1}{3}$; one third

3. $\dfrac{1}{5} + \dfrac{1}{5} = \dfrac{2}{5}$

4. $\dfrac{2}{4} + \dfrac{2}{4} = \dfrac{4}{4}$

5. $\dfrac{2}{8} + \dfrac{5}{8} = \dfrac{7}{8}$

6. $\dfrac{4}{5} - \dfrac{1}{5} = \dfrac{3}{5}$

7. $\dfrac{4}{6} - \dfrac{2}{6} = \dfrac{2}{6}$

8. $\dfrac{9}{10} - \dfrac{4}{10} = \dfrac{5}{10}$

9. $12 \div 4 = 3$
$3 \times 3 = 9$

10. $40 \div 5 = 8$
$8 \times 2 = 16$

11. $14 \div 7 = 2$
$2 \times 5 = 10$

12. $6 + 6 + 10 = 22$ ft

13. $28 + 16 + 28 + 16 = 88$ yd

14. $17 + 17 + 17 + 17 = 68$ in

15. $\dfrac{3}{8} + \dfrac{4}{8} = \dfrac{7}{8}$ of the day

16. $\dfrac{2}{4} + \dfrac{1}{4} = \dfrac{3}{4}$ peanuts or almonds

$\dfrac{4}{4} - \dfrac{3}{4} = \dfrac{1}{4}$ not peanuts or almonds

17. 1 flower bed:
$6 + 8 + 12 = 26$ ft
$26 \times 2 = 52$ ft

18. $\dfrac{2}{5} + \dfrac{1}{5} = \dfrac{3}{5}$ of the apples

19. $10 \div 5 = 2$

$2 \times 3 = 6$ eaten

$10 - 6 = 4$ left

20. $\$3.15 + \$0.95 + \$1.25 = \$5.35$

Systematic Review 3F

1. $\dfrac{3}{6} + \dfrac{2}{6} = \dfrac{5}{6}$ five sixths

2. $\dfrac{4}{4} - \dfrac{2}{4} = \dfrac{2}{4}$ two fourths

3. $\dfrac{1}{3} + \dfrac{1}{3} = \dfrac{2}{3}$

4. $\dfrac{2}{5} + \dfrac{1}{5} = \dfrac{3}{5}$

5. $\dfrac{3}{11} + \dfrac{4}{11} = \dfrac{7}{11}$

6. $\dfrac{5}{9} - \dfrac{4}{9} = \dfrac{1}{9}$

7. $\dfrac{6}{7} - \dfrac{1}{7} = \dfrac{5}{7}$

8. $\dfrac{3}{3} - \dfrac{1}{3} = \dfrac{2}{3}$

9. $28 \div 7 = 4$

$4 \times 3 = 12$

10. $27 \div 3 = 9$

$9 \times 1 = 9$

11. $36 \div 4 = 9$

$9 \times 3 = 27$

12. $63 + 87 + 48 = 198$ ft

13. $35 + 18 + 35 + 18 = 106$ yd

14. $81 + 81 + 81 + 81 = 324$ in

15. $\dfrac{1}{5} + \dfrac{3}{5} = \dfrac{4}{5}$ of a foot

16. $10 \div 5 = 2$

$2 \times 4 = 8$ pizzas eaten

$10 - 8 = 2$ pizzas left over

17. $\dfrac{3}{3} = \dfrac{1}{3} = \dfrac{2}{3}$ of a sandwich

18. $16 + 16 + 16 = 48$ ft

19. $\$17.45 - \$3.19 = \$14.26$

20. $12 \div 6 = 2$

$1 \times 2 = 2$ months

Lesson Practice 4A

1. done

2. $\dfrac{1}{5} = \dfrac{2}{10} = \dfrac{3}{15} = \dfrac{4}{20} = \dfrac{5}{25}$

one fifth = two tenths = three fifteenths = four twentieths = five twenty-fifths

3. $\dfrac{2}{3} = \dfrac{4}{6} = \dfrac{6}{9} = \dfrac{8}{12} = \dfrac{10}{15}$

two thirds = four sixths = six ninths = eight twelfths = ten fifteenths

4. $\dfrac{3}{6} = \dfrac{6}{12} = \dfrac{9}{18} = \dfrac{12}{24} = \dfrac{15}{30}$

three sixths = six twelfths = nine eighteenths = twelve twenty-fourths = fifteen thirtieths

5. done

6. $\dfrac{3}{5} = \dfrac{6}{10} = \dfrac{9}{15} = \dfrac{12}{20}$

Lesson Practice 4B

1. $\dfrac{3}{4} = \dfrac{6}{8} = \dfrac{9}{12} = \dfrac{12}{16} = \dfrac{15}{20}$

three fourths = six eighths = nine twelfths = twelve sixteenths = fifteen twentieths

2. $\dfrac{2}{2} = \dfrac{4}{4} = \dfrac{6}{6} = \dfrac{8}{8} = \dfrac{10}{10}$

two halves = four fourths = six sixths = eight eighths = ten tenths

3. $\dfrac{4}{5} = \dfrac{8}{10} = \dfrac{12}{15} = \dfrac{16}{20} = \dfrac{20}{25}$

four fifths = eight tenths = twelve fifteenths = sixteen twentieths = twenty twenty-fifths

4. $\dfrac{1}{3} = \dfrac{2}{6} = \dfrac{3}{9} = \dfrac{4}{12}$

5. $\dfrac{2}{4} = \dfrac{4}{8} = \dfrac{6}{12} = \dfrac{8}{16}$

6. $\dfrac{5}{6} = \dfrac{10}{12} = \dfrac{15}{18} = \dfrac{20}{24}$

7. $\dfrac{3}{3} = \dfrac{6}{6} = \dfrac{9}{9} = \dfrac{12}{12}$

8. $\dfrac{1}{3} = \dfrac{2}{6}$

Lesson Practice 4C

1. $\dfrac{1}{6} = \dfrac{2}{12} = \dfrac{3}{18} = \dfrac{4}{24} = \dfrac{5}{30}$

one sixth = two twelfths = three eighteenths = four twenty-fourths = five thirtieths

2. $\dfrac{2}{5} = \dfrac{4}{10} = \dfrac{6}{15} = \dfrac{8}{20} = \dfrac{10}{25}$

two fifths = four tenths = six fifteenths = eight twentieths = ten twenty-fifths

3. $\dfrac{5}{6} = \dfrac{10}{12} = \dfrac{15}{18} = \dfrac{20}{24} = \dfrac{25}{30}$

five sixths = ten twelfths = fifteen eighteenths = twenty twenty-fourths = twenty-five thirtieths

4. $\dfrac{3}{4} = \dfrac{6}{8} = \dfrac{9}{12} = \dfrac{12}{16}$

5. $\dfrac{1}{5} = \dfrac{2}{10} = \dfrac{3}{15} = \dfrac{4}{20}$

6. $\dfrac{4}{5} = \dfrac{8}{10} = \dfrac{12}{15} = \dfrac{16}{20}$

7. $\dfrac{2}{3} = \dfrac{4}{6} = \dfrac{6}{9} = \dfrac{8}{12}$

8. $\dfrac{1}{4} = \dfrac{2}{8}$

Systematic Review 4D

1. $\dfrac{4}{4} = \dfrac{8}{8} = \dfrac{12}{12} = \dfrac{16}{16} = \dfrac{20}{20}$

four fourths = eight eighths = twelve twelfths = sixteen sixteenths = twenty-twentieths

2. $\dfrac{1}{4} = \dfrac{2}{8} = \dfrac{3}{12} = \dfrac{4}{16}$

3. $\dfrac{1}{5} = \dfrac{2}{10} = \dfrac{3}{15} = \dfrac{4}{20}$

4. $\dfrac{2}{6} = \dfrac{4}{12} = \dfrac{6}{18} = \dfrac{8}{24}$

5. $\dfrac{3}{3} = \dfrac{6}{6} = \dfrac{9}{9} = \dfrac{12}{12}$

6. $\dfrac{3}{6} + \dfrac{2}{6} = \dfrac{5}{6}$

7. $\dfrac{5}{7} - \dfrac{3}{7} = \dfrac{2}{7}$

8. $\dfrac{1}{4} + \dfrac{1}{4} = \dfrac{2}{4}$

9. $36 \div 6 = 6$
$6 \times 1 = 6$

10. $32 \div 8 = 4$
$4 \times 5 = 20$

11. $81 \div 9 = 9$
$9 \times 3 = 27$

12. done

13. $16 \times 11 = 176$

14. $123 \times 32 = 3,936$

15. $2 : \dfrac{1}{3} \times \dfrac{2}{2} = \dfrac{2}{6}$

16. $\dfrac{1}{2} = \dfrac{4}{8}$: yes

17. $\dfrac{3}{4}$ of $36
$36 \div 4 = 9$
$9 \times 3 = 27

18. $\dfrac{2}{7} + \dfrac{3}{7} = \dfrac{5}{7}$
$\dfrac{7}{7} - \dfrac{5}{7} = \dfrac{2}{7}$ of the chores

Systematic Review 4E

1. $\dfrac{1}{3} = \dfrac{2}{6} = \dfrac{3}{9} = \dfrac{4}{12} = \dfrac{5}{15}$

one third = two sixths = three ninths = four twelfths = five fifteenths

2. $\dfrac{4}{6} = \dfrac{8}{12} = \dfrac{12}{18} = \dfrac{16}{24}$

3. $\dfrac{2}{5} = \dfrac{4}{10} = \dfrac{6}{15} = \dfrac{8}{20}$

4. $\dfrac{1}{2} = \dfrac{2}{4} = \dfrac{3}{6} = \dfrac{4}{8}$

5. $\dfrac{1}{3} = \dfrac{2}{6} = \dfrac{3}{9} = \dfrac{4}{12}$

6. $\dfrac{3}{4} + \dfrac{1}{4} = \dfrac{4}{4}$

7. $\dfrac{6}{9} - \dfrac{2}{9} = \dfrac{4}{9}$

8. $\dfrac{8}{8} - \dfrac{5}{8} = \dfrac{3}{8}$

9. $49 \div 7 = 7$
   $7 \times 4 = 28$

10. $45 \div 5 = 9$
    $9 \times 3 = 27$

11. $12 \div 2 = 6$
    $6 \times 1 = 6$

12. $22 \times 12 = 264$

13. $23 \times 13 = 299$

14. $405 \times 11 = 4,455$

15. $\dfrac{2}{3} = \dfrac{4}{6}$

16. $\dfrac{2}{3}$ of 18:
    $18 \div 3 = 6$
    $6 \times 2 = 12$ students

17. $\dfrac{1}{2} = \dfrac{2}{4}$

18. $\dfrac{1}{4} + \dfrac{2}{4} = \dfrac{3}{4}$ of a century

19. $\dfrac{6}{6} - \dfrac{1}{6} = \dfrac{5}{6}$ of the bushes

20. blue : $\dfrac{1}{6}$ of 12
    $12 \div 6 = 2 ; 2 \times 1 = 2$ bushes
    red : $12 - 2 = 10$ bushes

Systematic Review 4F

1. $\dfrac{1}{2} = \dfrac{2}{4} = \dfrac{3}{6} = \dfrac{4}{8} = \dfrac{5}{10}$
   one half = two fourths =
   three sixths = four
   eighths = five tenths

2. $\dfrac{1}{6} = \dfrac{2}{12} = \dfrac{3}{18} = \dfrac{4}{24}$

3. $\dfrac{3}{5} = \dfrac{6}{10} = \dfrac{9}{15} = \dfrac{12}{20}$

4. $\dfrac{1}{7} = \dfrac{2}{14} = \dfrac{3}{21} = \dfrac{4}{28}$

5. $\dfrac{3}{8} = \dfrac{6}{16} = \dfrac{9}{24} = \dfrac{12}{32}$

6. $\dfrac{2}{3} - \dfrac{1}{3} = \dfrac{1}{3}$

7. $\dfrac{3}{5} + \dfrac{1}{5} = \dfrac{4}{5}$

8. $\dfrac{3}{10} - \dfrac{2}{10} = \dfrac{1}{10}$

9. $60 \div 10 = 6$
   $6 \times 1 = 6$

10. $16 \div 8 = 2$
    $2 \times 2 = 4$

11. $9 \div 3 = 3$
    $3 \times 3 = 9$

12. $12 \times 11 = 132$

13. $14 \times 12 = 168$

14. $221 \times 43 = 9,503$

15. $\dfrac{1}{7} = \dfrac{2}{14}$

16. $\dfrac{2}{14}$ of 14 = 2 pounds
    $14 \div 14 = 1;$
    $1 \times 2 = 2$

17. $11 \times 13 = 143$ rolls

18. $4 + 4 + 4 + 4 = 16$ ft

19. $\dfrac{3}{8} + \dfrac{1}{8} = \dfrac{4}{8}$ done
    $\dfrac{8}{8} - \dfrac{4}{8} = \dfrac{4}{8}$ left to drive

20. $\dfrac{4}{8}$ of 64:
    $64 \div 8 = 8$
    $8 \times 4 = 32$ miles

Lesson Practice 5A

1. done

2. $\dfrac{1}{3} = \dfrac{2}{6} = \dfrac{3}{9} = \dfrac{4}{12} = \boxed{\dfrac{5}{15}}$
   $\dfrac{2}{5} = \dfrac{4}{10} = \boxed{\dfrac{6}{15}} = \dfrac{8}{20} = \dfrac{10}{25}$
   $\dfrac{5}{15} + \dfrac{6}{15} = \dfrac{11}{15}$

3. $\dfrac{1}{4} = \dfrac{2}{8} = \dfrac{3}{12} = \dfrac{4}{16} = \boxed{\dfrac{5}{20}}$
   $\dfrac{1}{5} = \dfrac{2}{10} = \dfrac{3}{15} = \boxed{\dfrac{4}{20}} = \dfrac{5}{25}$
   $\dfrac{5}{20} + \dfrac{4}{20} = \dfrac{9}{20}$

4. $\dfrac{1}{2} = \dfrac{2}{4} = \boxed{\dfrac{3}{6}} = \dfrac{4}{8} = \dfrac{5}{10}$
   $\boxed{\dfrac{1}{6}} = \dfrac{2}{12} = \dfrac{3}{18} = \dfrac{4}{24} = \dfrac{5}{30}$
   $\dfrac{3}{6} + \dfrac{1}{6} = \dfrac{4}{6}$

Lesson Practice 5B

1. $\dfrac{5}{6} = \boxed{\dfrac{10}{12}} = \dfrac{15}{18} = \dfrac{20}{24} = \dfrac{25}{30}$

   $\dfrac{3}{4} = \dfrac{6}{8} = \boxed{\dfrac{9}{12}} = \dfrac{12}{16} = \dfrac{15}{20}$

   $\dfrac{10}{12} - \dfrac{9}{12} = \dfrac{1}{12}$

2. $\dfrac{1}{2} = \dfrac{2}{4} = \boxed{\dfrac{3}{6}} = \dfrac{4}{8} = \dfrac{5}{10}$

   $\dfrac{1}{3} = \boxed{\dfrac{2}{6}} = \dfrac{3}{9} = \dfrac{4}{12} = \dfrac{5}{15}$

   $\dfrac{3}{6} - \dfrac{2}{6} = \dfrac{1}{6}$

3. $\dfrac{3}{4} = \dfrac{6}{8} = \boxed{\dfrac{9}{12}} = \dfrac{12}{16} = \dfrac{15}{20}$

   $\dfrac{2}{3} = \dfrac{4}{6} = \dfrac{6}{9} = \boxed{\dfrac{8}{12}} = \dfrac{10}{15}$

   $\dfrac{9}{12} - \dfrac{8}{12} = \dfrac{1}{12}$

4. $\dfrac{2}{4} = \dfrac{4}{8} = \dfrac{6}{12} = \dfrac{8}{16} = \boxed{\dfrac{10}{20}}$

   $\dfrac{1}{5} = \dfrac{2}{10} = \dfrac{3}{15} = \boxed{\dfrac{4}{20}} = \dfrac{5}{25}$

   $\dfrac{10}{20} - \dfrac{4}{20} = \dfrac{6}{20}$

Lesson Practice 5C

1. done

2. $\dfrac{12}{15} - \dfrac{5}{15} = \dfrac{7}{15}$

3. $\dfrac{10}{20} + \dfrac{8}{20} = \dfrac{18}{20}$

4. $\dfrac{6}{12} - \dfrac{4}{12} = \dfrac{2}{12}$

5. $\dfrac{6}{18} + \dfrac{9}{18} = \dfrac{15}{18}$

6. $\dfrac{9}{15} - \dfrac{5}{15} = \dfrac{4}{15}$

7. $\dfrac{10}{20} + \dfrac{4}{20} = \dfrac{14}{20}$

Systematic Review 5D

1. $\dfrac{3}{6} + \dfrac{2}{6} = \dfrac{5}{6}$

2. $\dfrac{5}{15} + \dfrac{6}{15} = \dfrac{11}{15}$

3. $\dfrac{12}{24} + \dfrac{6}{24} = \dfrac{18}{24}$

4. $\dfrac{9}{18} - \dfrac{6}{18} = \dfrac{3}{18}$

5. $\dfrac{12}{18} - \dfrac{3}{18} = \dfrac{9}{18}$

6. $\dfrac{6}{30} - \dfrac{5}{30} = \dfrac{1}{30}$

7. $\dfrac{2}{3} = \dfrac{4}{6} = \dfrac{6}{9} = \dfrac{8}{12}$

8. $\dfrac{4}{7} = \dfrac{8}{14} = \dfrac{12}{21} = \dfrac{16}{28}$

9. $24 \div 4 = 6$

   $6 \times 1 = 6$

10. $25 \div 5 = 5$

    $5 \times 2 = 10$

11. $63 \div 7 = 9$

    $9 \times 1 = 9$

12. 30

13. 80

14. 400

15. 700

16. $\dfrac{5}{6}$ of 60:

    $60 \div 6 = 10$

    $10 \times 5 = 50$ turkeys

17. $\dfrac{2}{3} + \dfrac{1}{6} = \dfrac{12}{18} + \dfrac{3}{18} = \dfrac{15}{18}$ of a pizza

18. $31 \times 12 = 372$ months

Systematic Review 5E

1. $\dfrac{6}{30} + \dfrac{5}{30} = \dfrac{11}{30}$

2. $\dfrac{5}{15} + \dfrac{9}{15} = \dfrac{14}{15}$

3. $\dfrac{5}{20} + \dfrac{4}{20} = \dfrac{9}{20}$

4. $\dfrac{25}{30} - \dfrac{6}{30} = \dfrac{19}{30}$

5. $\dfrac{16}{20} - \dfrac{15}{20} = \dfrac{1}{20}$

6. $\dfrac{6}{18} - \dfrac{3}{18} = \dfrac{3}{18}$

7. $\dfrac{1}{2} = \dfrac{2}{4} = \dfrac{3}{6} = \dfrac{4}{8}$

8. $\dfrac{3}{6} = \dfrac{6}{12} = \dfrac{9}{18} = \dfrac{12}{24}$

9. $18 \div 3 = 6$

   $6 \times 1 = 6$

10. $42 \div 7 = 6$
$6 \times 3 = 18$

11. $9 \div 9 = 1$
$1 \times 2 = 2$

12. 50

13. 60

14. 800

15. 100

16. $\frac{1}{3} + \frac{1}{2} = \frac{2}{6} + \frac{3}{6} = \frac{5}{6}$ of a mile

17. $\frac{2}{5} + \frac{1}{4} = \frac{8}{20} + \frac{5}{20} = \frac{13}{20}$
of the children

18. $\frac{2}{5}$ of 20:
$20 \div 5 = 4$
$4 \times 2 = 8$ children bought pears
$\frac{1}{4}$ of 20:
$20 \div 4 = 5$
$5 \times 1 = 5$ children bought apples

19. $221 \times 4 = 884$ dirty paws

20. $8 + 13 + 19 = 40$ ft

Systematic Review 5F

1. $\frac{12}{24} + \frac{8}{24} = \frac{20}{24}$

2. $\frac{5}{20} + \frac{12}{20} = \frac{17}{20}$

3. $\frac{6}{18} + \frac{6}{18} = \frac{12}{18}$

4. $\frac{3}{6} - \frac{2}{6} = \frac{1}{6}$

5. $\frac{8}{20} - \frac{5}{20} = \frac{3}{20}$

6. $\frac{15}{18} - \frac{6}{18} = \frac{9}{18}$

7. $\frac{3}{4} = \frac{6}{8} = \frac{9}{12} = \frac{12}{16}$

8. $\frac{5}{9} = \frac{10}{18} = \frac{15}{27} = \frac{20}{36}$

9. $25 \div 5 = 5$
$5 \times 2 = 10$

10. $30 \div 3 = 10$
$10 \times 2 = 20$

11. $56 \div 8 = 7$
$7 \times 1 = 7$

12. 60

13. 10

14. 100

15. 400

16. $\frac{3}{4} - \frac{1}{2} = \frac{6}{8} - \frac{4}{8} = \frac{2}{8}$ of a yard

17. $\frac{3}{8}$ of 80:
$80 \div 8 = 10$
$10 \times 3 = 30$ people with flags

18. $\$15.34 + \$19.99 + \$16.50$
$= \$51.83$

19. $50 + 30 + 50 + 30 = 160$ ft

20. $12 \times 24 = 288$ people

Lesson Practice 6A

1. done

2. $\frac{5}{10} + \frac{2}{10} = \frac{7}{10}$

3. $\frac{4}{12} + \frac{6}{12} = \frac{10}{12}$

4. $\frac{12}{18} - \frac{6}{18} = \frac{6}{18}$

5. $\frac{12}{30} - \frac{5}{30} = \frac{7}{30}$

6. $\frac{5}{20} - \frac{4}{20} = \frac{1}{20}$

Lesson Practice 6B

1. done

2. $\frac{5}{20} + \frac{12}{20} = \frac{17}{20}$

3. $\frac{3}{18} + \frac{6}{18} = \frac{9}{18}$

4. $\frac{8}{12} + \frac{3}{12} = \frac{11}{12}$

5. $\frac{20}{30} - \frac{6}{30} = \frac{14}{30}$

6. $\frac{4}{8} - \frac{2}{8} = \frac{2}{8}$

7. $\frac{12}{15} - \frac{10}{15} = \frac{2}{15}$

8. $\frac{6}{12} - \frac{4}{12} = \frac{2}{12}$

9. $\frac{3}{4}+\frac{1}{6}=\frac{18}{24}+\frac{4}{24}=\frac{22}{24}$
of the casserole

10. $\frac{3}{4}-\frac{1}{2}=\frac{6}{8}-\frac{4}{8}=\frac{2}{8}$ in

Lesson Practice 6C

1. $\frac{4}{12}+\frac{3}{12}=\frac{7}{12}$

2. $\frac{5}{35}+\frac{14}{35}=\frac{19}{35}$

3. $\frac{6}{16}+\frac{8}{16}=\frac{14}{16}$

4. $\frac{4}{24}+\frac{18}{24}=\frac{22}{24}$

5. $\frac{25}{30}-\frac{18}{30}=\frac{7}{30}$

6. $\frac{16}{20}-\frac{5}{20}=\frac{11}{20}$

7. $\frac{5}{15}-\frac{3}{15}=\frac{2}{15}$

8. $\frac{14}{18}-\frac{9}{18}=\frac{5}{18}$

9. $\frac{1}{2}+\frac{1}{4}=\frac{4}{8}+\frac{2}{8}=\frac{6}{8}$
of her books

10. $\frac{2}{5}-\frac{1}{3}=\frac{6}{15}-\frac{5}{15}=\frac{1}{15}$
of a bucket

Systematic Review 6D

1. $\frac{15}{20}+\frac{4}{20}=\frac{19}{20}$

2. $\frac{6}{24}+\frac{8}{24}=\frac{14}{24}$

3. $\frac{12}{18}+\frac{3}{18}=\frac{15}{18}$

4. $\frac{6}{24}-\frac{4}{24}=\frac{2}{24}$

5. $\frac{10}{30}-\frac{6}{30}=\frac{4}{30}$

6. $\frac{32}{72}-\frac{9}{72}=\frac{23}{72}$

7. $\frac{1}{5}=\frac{2}{10}=\frac{3}{15}=\frac{4}{20}$

8. $\frac{4}{5}=\frac{8}{10}=\frac{12}{15}=\frac{16}{20}$

9. $12\div3=4$
$4\times1=4$

10. $16\div8=2$
$2\times7=14$

11. $81\div9=9$
$9\times4=36$

12. $(20)\times(40)=(800)$
$23\times36=828$

13. $(80)\times(30)=(2,400)$
$78\times34=2,652$

14. $(70)\times(20)=(1,400)$
$65\times15=975$

15. $15\times24=360$ tomatoes

16. $\frac{2}{3}$ of $9 = $6 on hand
$9 - $6 = $3 needed

17. $\frac{1}{5}+\frac{1}{6}=\frac{6}{30}+\frac{5}{30}=\frac{11}{30}$ of the troops

18. $\frac{1}{2}+\frac{2}{5}=\frac{5}{10}+\frac{4}{10}=\frac{9}{10}$ in

Systematic Review 6E

1. $\frac{6}{30}+\frac{15}{30}=\frac{21}{30}$

2. $\frac{3}{21}+\frac{14}{21}=\frac{17}{21}$

3. $\frac{5}{15}+\frac{3}{15}=\frac{8}{15}$

4. $\frac{15}{18}-\frac{12}{18}=\frac{3}{18}$

5. $\frac{4}{12}-\frac{3}{12}=\frac{1}{12}$

6. $\frac{9}{18}-\frac{2}{18}=\frac{7}{18}$

7. $\frac{3}{7}=\frac{6}{14}=\frac{9}{21}=\frac{12}{28}$

8. $\frac{9}{11}=\frac{18}{22}=\frac{27}{33}=\frac{36}{44}$

9. $35\div7=5$
$5\times1=5$

10. $50\div5=10$
$10\times4=40$

11. $36\div4=9$
$9\times3=27$

12. $(50)\times(20)=(1,000)$
$45\times24=1,080$

13. $(70) \times (20) = (1,400)$
$67 \times 18 = 1,206$

14. $(30) \times (40) = (1,200)$
$32 \times 39 = 1,248$

15. $15 \times \$33 = \$495$

16. $80 \div 10 = 8$
$8 \times 9 = 72$ posts

17. $\dfrac{3}{6} + \dfrac{1}{6} = \dfrac{4}{6}$ gone

$\dfrac{6}{6} - \dfrac{4}{6} = \dfrac{2}{6}$ left

18. $\dfrac{1}{6} - \dfrac{1}{12} = \dfrac{12}{72} - \dfrac{6}{72} = \dfrac{6}{72}$ ft

19. 500

20. $6 + 6 + 6 + 6 = 24$ yd

Systematic Review 6F

1. $\dfrac{20}{30} + \dfrac{6}{30} = \dfrac{26}{30}$

2. $\dfrac{6}{12} + \dfrac{4}{12} = \dfrac{10}{12}$

3. $\dfrac{22}{77} + \dfrac{21}{77} = \dfrac{43}{77}$

4. $\dfrac{5}{10} - \dfrac{2}{10} = \dfrac{3}{10}$

5. $\dfrac{8}{12} - \dfrac{3}{12} = \dfrac{5}{12}$

6. $\dfrac{18}{72} - \dfrac{16}{72} = \dfrac{2}{72}$

7. $\dfrac{5}{6} = \dfrac{10}{12} = \dfrac{15}{18} = \dfrac{20}{24}$

8. $\dfrac{1}{10} = \dfrac{2}{20} = \dfrac{3}{30} = \dfrac{4}{40}$

9. $28 \div 7 = 4$;
$4 \times 3 = 12$

10. $54 \div 6 = 9$; $9 \times 1 = 9$

11. $8 \div 8 = 1$; $1 \times 4 = 4$

12. $(70) \times (90) = (6,300)$
$73 \times 89 = 6,497$

13. $(30) \times (90) = (2,700)$
$26 \times 91 = 2,366$

14. $(50) \times (10) = (500)$
$47 \times 11 = 517$

15. $18 \times 12 = 216$ eggs

16. $312 \times 3 = 936$ mi

17. $\dfrac{4}{9} + \dfrac{3}{6} = \dfrac{24}{54} + \dfrac{27}{54} = \dfrac{51}{54}$ of a loaf

18. $60 \div 6 = 10$
$5 \times 10 = 50$ min

19. 300

20. $13 + 18 + 13 + 18 = 62$ in
$2 \times 24 = 48$ in
62 in > 48 in; no

Lesson Practice 7A

1. done

2. $\dfrac{9}{15} < \dfrac{10}{15}$, so $\dfrac{3}{5} < \dfrac{2}{3}$

3. done

4. $\dfrac{18}{30} < \dfrac{20}{30}$, so $\dfrac{3}{5} < \dfrac{4}{6}$

5. $\dfrac{8}{12} < \dfrac{9}{12}$, so $\dfrac{2}{3} < \dfrac{3}{4}$

6. $\dfrac{6}{15} > \dfrac{5}{15}$, so $\dfrac{2}{5} > \dfrac{1}{3}$

Lesson Practice 7B

1. done

2. $\dfrac{12}{18} < \dfrac{15}{18}$, so $\dfrac{2}{3} < \dfrac{5}{6}$
less than

3. $\dfrac{12}{18} > \dfrac{9}{18}$, so $\dfrac{2}{3} > \dfrac{3}{6}$
greater than

4. $\dfrac{5}{10} > \dfrac{4}{10}$, so $\dfrac{1}{2} > \dfrac{2}{5}$
greater than

5. $\dfrac{6}{18} = \dfrac{6}{18}$, so $\dfrac{1}{3} = \dfrac{2}{6}$
equal

6. $\dfrac{10}{20} > \dfrac{4}{20}$, so $\dfrac{2}{4} > \dfrac{1}{5}$
greater than

7. $\dfrac{5}{10} > \dfrac{4}{10}$, so $\dfrac{1}{2} > \dfrac{2}{5}$
Trisha received more votes.

8. $\dfrac{6}{18} < \dfrac{12}{18}$, so $\dfrac{2}{6} < \dfrac{2}{3}$
Donald ran farther.

Lesson Practice 7C

1. $\frac{24}{30} > \frac{20}{30}$, so $\frac{4}{5} > \frac{4}{6}$
   greater than

2. $\frac{8}{12} < \frac{12}{12}$, so $\frac{4}{6} < \frac{2}{2}$
   less than

3. $\frac{21}{56} < \frac{32}{56}$, so $\frac{3}{8} < \frac{4}{7}$
   less than

4. $\frac{6}{27} < \frac{9}{27}$, so $\frac{2}{9} < \frac{1}{3}$
   less than

5. $\frac{18}{24} < \frac{20}{24}$

6. $\frac{12}{24} = \frac{12}{24}$

7. $\frac{10}{20} > \frac{6}{20}$

8. $\frac{28}{35} < \frac{30}{35}$

9. $\frac{6}{24} > \frac{4}{24}$ Shirley ate more pizza.

10. $\frac{36}{60} > \frac{35}{60}$ The land on the east side is larger.

Systematic Review 7D

1. $\frac{6}{18} < \frac{9}{18}$

2. $\frac{10}{16} > \frac{8}{16}$

3. $\frac{12}{48} = \frac{12}{48}$

4. $\frac{12}{24} + \frac{4}{24} = \frac{16}{24}$

5. $\frac{48}{80} - \frac{30}{80} = \frac{18}{80}$

6. $\frac{14}{63} + \frac{45}{63} = \frac{59}{63}$

7. $\frac{6}{8} = \frac{12}{16} = \frac{18}{24} = \frac{24}{32}$

8. $6 \div 2 = 3$;
   $3 \times 1 = 3$

9. $42 \div 6 = 7$;
   $7 \times 3 = 21$

10. $24 \div 8 = 3$;
    $3 \times 3 = 9$

11. done

12. $23 \div 5 = 4\frac{3}{5}$

13. $59 \div 7 = 8\frac{3}{7}$

14. $17 \div 4 = 4\frac{1}{4}$ yd

15. $\frac{16}{56} < \frac{35}{56}$ Penny did the most.
    $\frac{16}{56} + \frac{35}{56} = \frac{51}{56}$ done
    $\frac{56}{56} - \frac{51}{56} = \frac{5}{56}$ left

16. $\frac{5}{56}$ of 56;
    $56 \div 56 = 1$; $1 \times 5 = 5$ chores

17. $\frac{3}{12} + \frac{4}{12} = \frac{7}{12}$ of a cup

18. $\frac{21}{24} - \frac{8}{24} = \frac{13}{24}$ in

Systematic Review 7E

1. $\frac{9}{15} > \frac{5}{15}$

2. $\frac{12}{18} > \frac{3}{18}$

3. $\frac{108}{120} > \frac{70}{120}$

4. $\frac{5}{10} + \frac{4}{10} = \frac{9}{10}$

5. $\frac{6}{12} - \frac{4}{12} = \frac{2}{12}$

6. $\frac{15}{40} + \frac{24}{40} = \frac{39}{40}$

7. $\frac{1}{10} = \frac{2}{20} = \frac{3}{30} = \frac{4}{40}$

8. $32 \div 8 = 4$; $4 \times 7 = 28$

9. $21 \div 7 = 3$; $3 \times 2 = 6$

10. $20 \div 4 = 5$; $5 \times 3 = 15$

11. $32 \div 6 = 5\frac{2}{6}$

12. $19 \div 8 = 2\frac{3}{8}$

13. $48 \div 5 = 9\frac{3}{5}$

14. $(20) \times (20) = (400)$
    $21 \times 16 = 336$

15. $(30) \times (30) = (900)$
    $34 \times 29 = 986$

16. $(80) \times (10) = (800)$
    $75 \times 12 = 900$
17. $\frac{7}{42} + \frac{6}{42} = \frac{13}{42}$ of the cars
18. $28 \div 7 = 4$
    $4 \times 4 = 16$ games
19. $8 + 10 + 8 + 10 = 36$
    $\frac{1}{4}$ of $36 = 9$ yd
20. $\frac{8}{16} > \frac{6}{16}$, so $\frac{1}{2} > \frac{3}{8}$

19. $30 \div 6 = 5$
    $5 \times 5 = 25$ questions for Kiley
    $30 \div 5 = 6$
    $6 \times 4 = 24$ questions for Casey
    $25 > 24$; yes
20. $\frac{20}{32} < \frac{24}{32}$, so $\frac{5}{8} < \frac{3}{4}$

Systematic Review 7F

1. $\frac{60}{120} = \frac{60}{120}$
2. $\frac{10}{35} < \frac{21}{35}$
3. $\frac{3}{6} < \frac{4}{6}$
4. $\frac{10}{15} + \frac{3}{15} = \frac{13}{15}$
5. $\frac{16}{24} - \frac{6}{24} = \frac{10}{24}$
6. $\frac{45}{54} + \frac{6}{54} = \frac{51}{54}$
7. $\frac{3}{4} = \frac{6}{8} = \frac{9}{12} = \frac{12}{16}$
8. $10 \div 5 = 2$; $2 \times 3 = 6$
9. $12 \div 4 = 3$; $3 \times 1 = 3$
10. $24 \div 6 = 4$; $4 \times 4 = 16$
11. $13 \div 3 = 4\frac{1}{3}$
12. $39 \div 4 = 9\frac{3}{4}$
13. $58 \div 9 = 6\frac{4}{9}$
14. $(60) \times (50) = (3,000)$
    $64 \times 51 = 3,264$
15. $(50) \times (20) = (1,000)$
    $45 \times 19 = 855$
16. $(80) \times (40) = (3,200)$
    $82 \times 37 = 3,034$
17. $8 + 9 + 10 = 27$ ft
18. $\frac{25}{30} > \frac{24}{30}$ so $\frac{5}{6} > \frac{4}{5}$

Lesson Practice 8A

1. done
2. $\frac{6}{30} + \frac{5}{30} = \frac{11}{30}$
   $\frac{11}{30} + \frac{1}{2} = \frac{22}{60} + \frac{30}{60} = \frac{52}{60}$
3. $\frac{3}{24} + \frac{8}{24} = \frac{11}{24}$
   $\frac{11}{24} + \frac{1}{6} = \frac{66}{144} + \frac{24}{144} = \frac{90}{144}$
4. $\frac{6}{24} + \frac{20}{24} = \frac{26}{24}$
   $\frac{26}{24} + \frac{1}{3} = \frac{78}{72} + \frac{24}{72} = \frac{102}{72} = 1\frac{30}{72}$
   Note: at this point, the final step is optional when working with fractions greater than one.
5. done
6. done
7. $\frac{3 \times 6 \times 10}{7 \times 6 \times 10} + \frac{1 \times 7 \times 10}{6 \times 7 \times 10} + \frac{3 \times 7 \times 6}{10 \times 7 \times 6} =$
   $\frac{180}{420} + \frac{70}{420} + \frac{126}{420} = \frac{376}{420}$
8. $\frac{4 \times 2}{5 \times 2} + \frac{3}{10} + \frac{1 \times 5}{2 \times 5} =$
   $\frac{8}{10} + \frac{3}{10} + \frac{5}{10} = \frac{16}{10} = 1\frac{6}{10}$
9. $\frac{5}{8} + \frac{7}{8} + \frac{4}{8} = \frac{16}{8} = 2$ ft
10. $\frac{5}{40} + \frac{8}{40} = \frac{13}{40}$
    $\frac{13}{40} + \frac{1}{4} = \frac{52}{160} + \frac{40}{160} = \frac{92}{160}$
    or $\frac{13}{40} + \frac{10}{40} = \frac{23}{40}$ of a tank

The student may begin to use shortcuts that yield fractions different from the answers given. As long as the fraction is equivalent to the given answer, it is correct. For example, 2/5 is the same as 4/10.

Lesson Practice 8B

1. $\frac{5}{10} + \frac{4}{10} = \frac{9}{10}$

   $\frac{9}{10} + \frac{5}{6} = \frac{54}{60} + \frac{50}{60} = \frac{104}{60} = 1\frac{44}{60}$

2. $\frac{10}{35} + \frac{7}{35} = \frac{17}{35}$

   $\frac{17}{35} + \frac{2}{3} = \frac{51}{105} + \frac{70}{105} = \frac{121}{105} = 1\frac{16}{105}$

3. $\frac{6}{16} + \frac{8}{16} = \frac{14}{16}$

   $\frac{14}{16} + \frac{2}{5} = \frac{70}{80} + \frac{32}{80} = \frac{102}{80} = 1\frac{22}{80}$

4. $\frac{9}{63} + \frac{14}{63} = \frac{23}{63}$

   $\frac{23}{63} + \frac{1}{3} = \frac{69}{189} + \frac{63}{189} = \frac{132}{189}$

5. $\frac{5 \times 2 \times 3}{7 \times 2 \times 3} + \frac{1 \times 7 \times 3}{2 \times 7 \times 3} + \frac{1 \times 7 \times 2}{3 \times 7 \times 2} =$

   $\frac{30}{42} + \frac{21}{42} + \frac{14}{42} = \frac{65}{42} = 1\frac{23}{42}$

6. $\frac{4}{8} + \frac{6}{8} + \frac{5}{8} = \frac{15}{8} = 1\frac{7}{8}$

7. $\frac{2 \times 4 \times 6}{5 \times 4 \times 6} + \frac{1 \times 5 \times 6}{4 \times 5 \times 6} + \frac{1 \times 5 \times 4}{6 \times 5 \times 4} =$

   $\frac{48}{120} + \frac{30}{120} + \frac{20}{120} = \frac{98}{120}$

8. $\frac{4}{9} + \frac{3}{9} + \frac{2}{9} = \frac{9}{9} = 1$

9. $\frac{1}{10} + \frac{4}{10} + \frac{5}{10} = \frac{10}{10} = 1$ mile

10. $\frac{5}{6} + \frac{3}{6} + \frac{4}{6} = \frac{12}{6} = 2$ oranges

Lesson Practice 8C

1. $\frac{6}{10} + \frac{5}{10} = \frac{11}{10}$

   $\frac{11}{10} + \frac{1}{3} = \frac{33}{30} + \frac{10}{30} = \frac{43}{30} = 1\frac{13}{30}$

2. $\frac{15}{18} + \frac{6}{18} = \frac{21}{18}$

   $\frac{21}{18} + \frac{2}{5} = \frac{105}{90} + \frac{36}{90} = \frac{141}{90}$

   $= 1\frac{51}{90}$ or $\frac{47}{30} = 1\frac{17}{30}$

3. $\frac{4}{12} + \frac{9}{12} = \frac{13}{12}$

   $\frac{13}{12} + \frac{5}{6} = \frac{78}{72} + \frac{60}{72} = \frac{138}{72}$

   $= 1\frac{66}{72}$ or $\frac{46}{24} = 1\frac{22}{24}$

4. $\frac{6}{18} + \frac{6}{18} = \frac{12}{18}$

   $\frac{12}{18} + \frac{1}{2} = \frac{24}{36} + \frac{18}{36} = \frac{42}{36} = 1\frac{6}{36}$ or $\frac{7}{6} = 1\frac{1}{6}$

5. $\frac{1}{5} + \frac{1}{5} = \frac{2}{5}$

   $\frac{2}{5} + \frac{3}{10} = \frac{20}{50} + \frac{15}{50} = \frac{35}{50}$ or $\frac{7}{10}$

6. $\frac{8}{20} + \frac{5}{20} = \frac{13}{20}$

   $\frac{13}{20} + \frac{1}{6} = \frac{78}{120} + \frac{20}{120} = \frac{98}{120}$

7. $\frac{12}{14} + \frac{3}{14} + \frac{7}{14} = \frac{22}{14} = 1\frac{8}{14}$ or $\frac{308}{196} = 1\frac{112}{196}$

8. $\frac{28}{32} + \frac{8}{32} = \frac{36}{32}$

   $\frac{36}{32} + \frac{1}{3} = \frac{108}{96} + \frac{32}{96} = \frac{140}{96} = 1\frac{44}{96}$

9. $\frac{1}{4} + \frac{1}{4} = \frac{2}{4}$

   $\frac{2}{4} + \frac{1}{8} = \frac{16}{32} + \frac{4}{32} = \frac{20}{32}$ or $\frac{5}{8}$ of the job

10. $\frac{4}{8} + \frac{2}{8} + \frac{1}{8} = \frac{7}{8}$ or $\frac{56}{64}$ of her money

Systematic Review 8D

1. $\frac{8}{20} + \frac{5}{20} = \frac{13}{20}$

   $\frac{13}{20} + \frac{1}{2} = \frac{26}{40} + \frac{20}{40} = \frac{46}{40}$

   $= 1\frac{6}{40}$ or $\frac{23}{20} = 1\frac{3}{20}$

2. $\frac{18}{20} + \frac{10}{20} = \frac{28}{20}$

   $\frac{28}{20} + \frac{3}{4} = \frac{112}{80} + \frac{60}{80} = \frac{172}{80}$

   $= 2\frac{12}{80}$ or $\frac{86}{40} = 2\frac{6}{40}$

3. $\frac{1}{6} + \frac{4}{6} + \frac{3}{6} = \frac{8}{6} = 1\frac{2}{6}$ or $\frac{48}{36} = 1\frac{12}{36}$

4. $\frac{6}{24} > \frac{4}{24}$

5. $\dfrac{9}{15} > \dfrac{5}{15}$

6. $\dfrac{12}{24} = \dfrac{12}{24}$

7. $\dfrac{2}{3} = \dfrac{4}{6} = \dfrac{6}{9} = \dfrac{8}{12}$

8. $\dfrac{1}{6} = \dfrac{2}{12} = \dfrac{3}{18} = \dfrac{4}{24}$

9. $73 \times 62 = 4,526$

10. $54 \times 28 = 1,512$

11. $91 \times 49 = 4,459$

12. done

13. $379 \div 6 = 63\dfrac{1}{6}$

14. $503 \div 2 = 251\dfrac{1}{2}$

15. $\dfrac{8}{16} + \dfrac{2}{16} + \dfrac{1}{16} = \dfrac{11}{16}$

$\dfrac{11}{16} < \dfrac{16}{16}$ Something else was used.

16. $35 \div 4 = 8$ r.3; 8 groups, 3 left over

17. $\dfrac{3}{4}$ of $8 = 6$ books

18. $\dfrac{7}{42} < \dfrac{12}{42}$ so $\dfrac{1}{6} < \dfrac{2}{7}$

Clara received more.

Systematic Review 8E

1. $\dfrac{20}{32} + \dfrac{8}{32} = \dfrac{28}{32}$

$\dfrac{28}{32} + \dfrac{1}{2} = \dfrac{56}{64} + \dfrac{32}{64} = \dfrac{88}{64}$

$= 1\dfrac{24}{64}$ or $\dfrac{11}{8} = 1\dfrac{3}{8}$

2. $\dfrac{4}{20} + \dfrac{10}{20} = \dfrac{14}{20}$

$\dfrac{14}{20} + \dfrac{5}{6} = \dfrac{84}{120} + \dfrac{100}{120} = \dfrac{184}{120} = 1\dfrac{64}{120}$

3. $\dfrac{3}{6} + \dfrac{2}{6} = \dfrac{5}{6}$

$\dfrac{5}{6} + \dfrac{3}{6} = \dfrac{8}{6} = 1\dfrac{2}{6}$

4. $\dfrac{12}{18} > \dfrac{6}{18}$

5. $\dfrac{20}{32} > \dfrac{8}{32}$

6. $\dfrac{14}{35} < \dfrac{15}{35}$

7. $\dfrac{1}{8} = \dfrac{2}{16} = \dfrac{3}{24} = \dfrac{4}{32}$

8. $\dfrac{1}{2} = \dfrac{2}{4} = \dfrac{3}{6} = \dfrac{4}{8}$

9. $35 \times 22 = 770$

10. $47 \times 84 = 3,948$

11. $63 \times 19 = 1,197$

12. $198 \div 3 = 66$

13. $809 \div 8 = 101\dfrac{1}{8}$

14. $472 \div 4 = 118$

15. $\dfrac{3}{6} + \dfrac{2}{6} + \dfrac{1}{6} = \dfrac{6}{6} = 1$ pizza

16. $\dfrac{1}{2} = \dfrac{2}{4}$ of a pie

17. $45 \div 4 = 11$ r.1; 11 flashlights, 1 battery left

18. $122 \times 2 = 244$ hands

19. $85 \div 5 = 17$; $17 \times 2 = 34$ dimes

20. $\dfrac{3}{6} > \dfrac{2}{6}$ Jeremy grew more.

$\dfrac{3}{6} - \dfrac{2}{6} = \dfrac{1}{6}$ ft

Systematic Review 8F

1. $\dfrac{28}{40} + \dfrac{30}{40} = \dfrac{58}{40}$

$\dfrac{58}{40} + \dfrac{1}{3} = \dfrac{174}{120} + \dfrac{40}{120} = \dfrac{214}{120} = 1\dfrac{94}{120}$

Equivalent answers to addition and subtraction problems are correct.

2. $\dfrac{18}{21} + \dfrac{7}{21} = \dfrac{25}{21}$

$\dfrac{25}{21} + \dfrac{5}{6} = \dfrac{150}{126} + \dfrac{105}{126} = \dfrac{255}{126} = 2\dfrac{3}{126}$

3. $\dfrac{14}{16} + \dfrac{5}{16} + \dfrac{8}{16} = \dfrac{27}{16} = 1\dfrac{11}{16}$

4. $\dfrac{8}{18} < \dfrac{9}{18}$

5. $\dfrac{35}{42} < \dfrac{36}{42}$

6. $\dfrac{16}{24} > \dfrac{15}{24}$

7. $\dfrac{4}{7} = \dfrac{8}{14} = \dfrac{12}{21} = \dfrac{16}{28}$

8. $\dfrac{1}{3} = \dfrac{2}{6} = \dfrac{3}{9} = \dfrac{4}{12}$

9. $32 \times 55 = 1,760$

10. $76 \times 41 = 3,116$

11. $29 \times 17 = 493$

12. $361 \div 7 = 51\frac{4}{7}$

13. $734 \div 9 = 81\frac{5}{9}$

14. $108 \div 5 = 21\frac{3}{5}$

15. $\frac{2}{10} + \frac{1}{10} + \frac{5}{10} = \frac{8}{10}$

   $\frac{10}{10} - \frac{8}{10} = \frac{2}{10}$ of his allowance

16. 90

17. 100

18. $5 + 3 + 5 + 3 = 16$ mi

19. $28 \div 7 = 4$; $4 \times 1 = 4$ days

20. $25 \div 4 = 6\frac{1}{4}$ chocolates

Lesson Practice 9A

1. done

2. $\frac{3 \times 3}{5 \times 4} = \frac{9}{20}$

3. $\frac{1 \times 1}{3 \times 2} = \frac{1}{6}$

4. $\frac{4 \times 2}{6 \times 5} = \frac{8}{30}$

5. $\frac{2 \times 1}{3 \times 4} = \frac{2}{12}$

6. $\frac{1 \times 5}{5 \times 6} = \frac{5}{30}$

7. $\frac{4 \times 2}{5 \times 6} = \frac{8}{30}$

8. $\frac{1 \times 1}{6 \times 3} = \frac{1}{18}$

9. $\frac{1 \times 2}{2 \times 6} = \frac{2}{12}$

10. $\frac{3}{5} \times \frac{2}{5} = \frac{6}{25}$

11. $\frac{3}{6} \times \frac{2}{3} = \frac{6}{18}$

12. $\frac{2}{4} \times \frac{1}{5} = \frac{2}{20}$

13. $\frac{3}{4} \times \frac{4}{6} = \frac{12}{24}$

14. $\frac{2}{5} \times \frac{1}{2} = \frac{2}{10}$

15. $\frac{4}{6} \times \frac{1}{3} = \frac{4}{18}$

16. $\frac{1}{3} \times \frac{1}{4} = \frac{1}{12}$ of a cup

17. $\frac{1}{2} \times \frac{1}{5} = \frac{1}{10}$ of the group

18. $\frac{1}{3} \times \frac{1}{5} = \frac{1}{15}$ of a pie

Lesson Practice 9B

1. $\frac{2 \times 3}{6 \times 5} = \frac{6}{30}$

2. $\frac{2 \times 1}{3 \times 2} = \frac{2}{6}$

3. $\frac{3 \times 2}{8 \times 4} = \frac{6}{32}$

4. $\frac{1 \times 1}{4 \times 5} = \frac{1}{20}$

5. $\frac{1 \times 2}{6 \times 3} = \frac{2}{18}$

6. $\frac{1 \times 4}{3 \times 5} = \frac{4}{15}$

7. $\frac{4 \times 3}{5 \times 6} = \frac{12}{30}$

8. $\frac{1 \times 5}{2 \times 6} = \frac{5}{12}$

9. $\frac{3 \times 3}{4 \times 4} = \frac{9}{16}$

10. $\frac{2}{5} \times \frac{4}{6} = \frac{8}{30}$

11. $\frac{1}{7} \times \frac{1}{2} = \frac{1}{14}$

12. $\frac{3}{5} \times \frac{5}{6} = \frac{15}{30}$

13. $\frac{3}{10} \times \frac{1}{4} = \frac{3}{40}$

14. $\frac{1}{4} \times \frac{1}{2} = \frac{1}{8}$

15. $\frac{7}{8} \times \frac{1}{9} = \frac{7}{72}$

16. $\frac{3}{4} \times \frac{1}{4} = \frac{3}{16}$ of the class

17. $\frac{2}{3} \times \frac{3}{5} = \frac{6}{15}$ of a bushel

18. $\frac{1}{5} \times \frac{1}{5} = \frac{1}{25}$ of the games

Lesson Practice 9C

1. $\dfrac{3 \times 2}{6 \times 5} = \dfrac{6}{30}$

2. $\dfrac{2 \times 2}{9 \times 3} = \dfrac{4}{27}$

3. $\dfrac{2 \times 1}{6 \times 3} = \dfrac{2}{18}$

4. $\dfrac{3 \times 4}{4 \times 5} = \dfrac{12}{20}$

5. $\dfrac{1 \times 3}{7 \times 4} = \dfrac{3}{28}$

6. $\dfrac{1 \times 1}{6 \times 6} = \dfrac{1}{36}$

7. $\dfrac{4 \times 1}{8 \times 3} = \dfrac{4}{24}$

8. $\dfrac{3 \times 3}{5 \times 4} = \dfrac{9}{20}$

9. $\dfrac{4 \times 2}{6 \times 4} = \dfrac{8}{24}$

10. $\dfrac{1 \times 2}{6 \times 5} = \dfrac{2}{30}$

11. $\dfrac{5}{9} \times \dfrac{1}{4} = \dfrac{5}{36}$

12. $\dfrac{1 \times 1}{10 \times 10} = \dfrac{1}{100}$

13. $\dfrac{2 \times 3}{3 \times 6} = \dfrac{6}{18}$

14. $\dfrac{3 \times 1}{5 \times 2} = \dfrac{3}{10}$

15. $\dfrac{1 \times 3}{8 \times 5} = \dfrac{3}{40}$

16. $\dfrac{4}{5} \times \dfrac{2}{4} = \dfrac{8}{20}$ mi

17. $\dfrac{2}{3} \times \dfrac{1}{3} = \dfrac{2}{9}$ of a cup

18. $\dfrac{7}{8} \times \dfrac{1}{7} = \dfrac{7}{56}$ of a pie

Systematic Review 9D

1. $\dfrac{2}{5} \times \dfrac{1}{5} = \dfrac{2}{25}$

2. $\dfrac{5}{6} \times \dfrac{1}{4} = \dfrac{5}{24}$

3. $\dfrac{5}{9} \times \dfrac{1}{2} = \dfrac{5}{18}$

4. $\dfrac{3}{21} + \dfrac{7}{21} = \dfrac{10}{21}$

5. $\dfrac{14}{21} - \dfrac{3}{21} = \dfrac{11}{21}$

6. $\dfrac{5}{10} + \dfrac{4}{10} + \dfrac{7}{10} = \dfrac{16}{10}$ or $1\dfrac{6}{10}$

7. $\dfrac{28}{42} > \dfrac{6}{42}$

8. $\dfrac{16}{24} > \dfrac{15}{24}$

9. $\dfrac{11}{99} < \dfrac{18}{99}$

10. $\dfrac{2}{7} = \dfrac{4}{14} = \dfrac{6}{21} = \dfrac{8}{28}$

11. $\dfrac{1}{9} = \dfrac{2}{18} = \dfrac{3}{27} = \dfrac{4}{36}$

12. done

13. $(200) \times (60) = (12{,}000)$

    $179 \times 57 = 10{,}203$

14. $(900) \times (10) = (9{,}000)$

    $902 \times 11 = 9{,}922$

15. $\dfrac{1}{3} \times \dfrac{2}{3} = \dfrac{2}{9}$ of a cup

16. $365 \times 25 = 9{,}125$ days

17. $\dfrac{3}{6} + \dfrac{2}{6} = \dfrac{5}{6}$

    $\dfrac{5}{6} + \dfrac{1}{4} = \dfrac{20}{24} + \dfrac{6}{24} = \dfrac{26}{24}$

    of $1\dfrac{2}{24}$ cups

18. $36 \div 4 = 9$

    $9 \times 3 = 27$ hawks

Systematic Review 9E

1. $\dfrac{1}{2} \times \dfrac{2}{3} = \dfrac{2}{6}$

2. $\dfrac{7}{8} \times \dfrac{2}{5} = \dfrac{14}{40}$

3. $\dfrac{1}{3} \times \dfrac{3}{5} = \dfrac{3}{15}$

4. $\dfrac{27}{36} + \dfrac{4}{36} = \dfrac{31}{36}$

5. $\dfrac{10}{15} - \dfrac{6}{15} = \dfrac{4}{15}$

6. $\dfrac{3}{8} + \dfrac{5}{8} = \dfrac{8}{8} = 1$

    $1 + \dfrac{2}{8} = 1\dfrac{2}{8}$ or $1\dfrac{1}{4}$

7. $\dfrac{9}{27} = \dfrac{9}{27}$

8. $\dfrac{16}{40} > \dfrac{15}{40}$

9. $\dfrac{50}{70} > \dfrac{49}{70}$

10. $(100) \times (50) = (5,000)$
    $125 \times 51 = 6,375$

11. $(300) \times (40) = (12,000)$
    $254 \times 35 = 8,890$

12. $(600) \times (30) = (18,000)$
    $563 \times 26 = 14,638$

13. $107 \div 6 = 17\dfrac{5}{6}$

14. $395 \div 8 = 49\dfrac{3}{8}$

15. $459 \div 2 = 229\dfrac{1}{2}$

16. $\dfrac{1}{2} \times \dfrac{4}{5} = \dfrac{4}{10}$ of the chores

17. $\dfrac{5}{30} + \dfrac{6}{30} = \dfrac{11}{30}$
    $\dfrac{11}{30} + \dfrac{1}{4} = \dfrac{44}{120} + \dfrac{30}{120} = \dfrac{74}{120}$
    of the job

18. $250 \times 51 = 12,750$ in

19. $12 + 17 + 28 = 57$ in

20. $235 \div 5 = 47$; $47 \times 3 = 141$
    jelly beans

Systematic Review 9F

1. $\dfrac{2}{4} \times \dfrac{2}{5} = \dfrac{4}{20}$

2. $\dfrac{1}{3} \times \dfrac{2}{6} = \dfrac{2}{18}$

3. $\dfrac{1}{2} \times \dfrac{4}{9} = \dfrac{4}{18}$

4. $\dfrac{10}{15} + \dfrac{3}{15} = \dfrac{13}{15}$

5. $\dfrac{16}{40} - \dfrac{15}{40} = \dfrac{1}{40}$

6. $\dfrac{5}{20} + \dfrac{12}{20} = \dfrac{17}{20}$
   $\dfrac{17}{20} + \dfrac{2}{3} = \dfrac{51}{60} + \dfrac{40}{60} = \dfrac{91}{60}$ or $1\dfrac{31}{60}$

7. $\dfrac{40}{48} > \dfrac{24}{48}$

8. $\dfrac{27}{45} > \dfrac{20}{45}$

9. $\dfrac{24}{32} = \dfrac{24}{32}$

10. $(600) \times (60) = (36,000)$
    $558 \times 62 = 34,596$

11. $(400) \times (80) = (32,000)$
    $407 \times 83 = 33,781$

12. $(300) \times (10) = (3,000)$
    $349 \times 12 = 4,188$

13. $128 \div 7 = 18\dfrac{2}{7}$

14. $471 \div 3 = 157$

15. $298 \div 5 = 59\dfrac{3}{5}$

16. $\dfrac{3}{1} \times \dfrac{1}{4} = \dfrac{3}{4}$ ft

17. $\dfrac{4}{5} - \dfrac{1}{3} = \dfrac{12}{15} - \dfrac{5}{15} = \dfrac{7}{15}$ mile

18. $\dfrac{1}{3} \times \dfrac{1}{2} = \dfrac{1}{6}$; $\dfrac{1}{6}$ of 12 = 2 people

19. $13 + 19 + 26 = 58$ in

20. $30 \div 2 = 15$
    $15 \times 1 = 15$ days

Lesson Practice 10A

1. done

2. done

3. $\dfrac{15}{24} \div \dfrac{16}{24} = \dfrac{15 \div 16}{1} = \dfrac{15}{16}$

4. $\dfrac{12}{16} \div \dfrac{8}{16} = \dfrac{12 \div 8}{1} = \dfrac{12}{8}$ or $1\dfrac{4}{8}$

5. $\dfrac{4}{8} \div \dfrac{2}{8} = \dfrac{4 \div 2}{1} = 2$

6. $\dfrac{1}{2} \div \dfrac{1}{8} = \dfrac{8}{16} \div \dfrac{2}{16} = \dfrac{8 \div 2}{1} = 4$ times

7. $\dfrac{2}{3} \div \dfrac{1}{6} = \dfrac{12}{18} \div \dfrac{3}{18} = \dfrac{12 \div 3}{1} = 4$ pieces

8. $\dfrac{6}{8} \div \dfrac{1}{4} = \dfrac{24}{32} \div \dfrac{8}{32} = \dfrac{24 \div 8}{1} = 3$ people

Lesson Practice 10B

1. $\dfrac{12}{20} \div \dfrac{5}{20} = \dfrac{12 \div 5}{1} = \dfrac{12}{5}$ or $2\dfrac{2}{5}$

2. $\dfrac{18}{24} \div \dfrac{4}{24} = \dfrac{18 \div 4}{1} = \dfrac{18}{4}$ or $4\dfrac{2}{4}$

3. $\dfrac{3}{12} \div \dfrac{4}{12} = \dfrac{3 \div 4}{1} = \dfrac{3}{4}$

4. $\dfrac{16}{24} \div \dfrac{15}{24} = \dfrac{16 \div 15}{1} = \dfrac{16}{15}$ or $1\dfrac{1}{15}$

5. $\dfrac{5}{20} \div \dfrac{16}{20} = \dfrac{5 \div 16}{1} = \dfrac{5}{16}$

6. $\dfrac{3}{4} \div \dfrac{1}{8} = \dfrac{24}{32} \div \dfrac{4}{32} = \dfrac{24 \div 4}{1} = 6$ times

7. $\dfrac{9}{10} \div \dfrac{1}{10} = \dfrac{9 \div 1}{1} = 9$ volunteers

8. $\dfrac{5}{16} \div \dfrac{1}{16} = \dfrac{5 \div 1}{1} = 5$ times

Lesson Practice 10C

1. $\dfrac{6}{15} \div \dfrac{10}{15} = \dfrac{6 \div 10}{1} = \dfrac{6}{10}$

2. $\dfrac{3}{6} \div \dfrac{2}{6} = \dfrac{3 \div 2}{1} = \dfrac{3}{2}$ or $1\dfrac{1}{2}$

3. $\dfrac{10}{15} \div \dfrac{12}{15} = \dfrac{10 \div 12}{1} = \dfrac{10}{12}$

4. $\dfrac{8}{16} \div \dfrac{8}{16} = \dfrac{8 \div 8}{1} = 1$

5. $\dfrac{5}{15} \div \dfrac{6}{15} = \dfrac{5 \div 6}{1} = \dfrac{5}{6}$

6. $\dfrac{1}{3} \div \dfrac{1}{9} = \dfrac{9}{27} \div \dfrac{3}{27} = \dfrac{9 \div 3}{1} = 3$ people

7. $\dfrac{4}{5} \div \dfrac{1}{10} = \dfrac{40}{50} \div \dfrac{5}{50} = \dfrac{40 \div 5}{1} = 8$ people

8. $\dfrac{1}{3} \div \dfrac{1}{6} = \dfrac{6}{18} \div \dfrac{3}{18} = \dfrac{6 \div 3}{1} = 2$ boards

Don't forget that the last step in these problems is optional for now.

Systematic Review 10D

1. $\dfrac{10}{12} \div \dfrac{6}{12} = \dfrac{10 \div 6}{1} = \dfrac{10}{6}$ or $1\dfrac{4}{6}$

2. $\dfrac{4}{5} \div \dfrac{2}{5} = \dfrac{4 \div 2}{1} = 2$

3. $\dfrac{12}{15} \div \dfrac{5}{15} = \dfrac{12 \div 5}{1} = \dfrac{12}{5}$ or $2\dfrac{2}{5}$

4. $\dfrac{3}{4} \times \dfrac{1}{8} = \dfrac{3}{32}$

5. $\dfrac{2}{3} \times \dfrac{1}{6} = \dfrac{2}{18}$

6. $\dfrac{6}{8} \times \dfrac{1}{4} = \dfrac{6}{32}$

7. $\dfrac{12}{18} - \dfrac{3}{18} = \dfrac{9}{18}$

8. $\dfrac{5}{10} - \dfrac{4}{10} = \dfrac{1}{10}$

9. $\dfrac{3}{6} + \dfrac{4}{6} = \dfrac{7}{6}$

$\dfrac{7}{6} + \dfrac{4}{5} = \dfrac{35}{30} + \dfrac{24}{30} = \dfrac{59}{30}$ or $1\dfrac{29}{30}$

10. $38 \times 94 = 3,572$

11. $237 \times 15 = 3,555$

12. $709 \times 51 = 36,159$

13. done

14. $(500) \div (50) = (10)$

15. $(600) \div (30) = (20)$

16. $\dfrac{1}{2} \times \dfrac{5}{8} = \dfrac{5}{16}$ of a pizza

17. $\dfrac{4}{6} \div \dfrac{1}{3} = \dfrac{12}{18} \div \dfrac{6}{18} = \dfrac{12 \div 6}{1} = 2$ pieces

18. $\dfrac{1}{2} + \dfrac{1}{6} = \dfrac{6}{12} + \dfrac{2}{12} = \dfrac{8}{12}$
of the choir

Systematic Review 10E

1. $\dfrac{2}{3} \div \dfrac{1}{3} = \dfrac{2 \div 1}{1} = 2$

2. $\dfrac{15}{20} \div \dfrac{8}{20} = \dfrac{15 \div 8}{1} = \dfrac{15}{8}$ or $1\dfrac{7}{8}$

3. $\dfrac{8}{12} \div \dfrac{6}{12} = \dfrac{8 \div 6}{1} = \dfrac{8}{6}$ or $1\dfrac{2}{6}$

4. $\dfrac{4}{5} \times \dfrac{2}{5} = \dfrac{8}{25}$

5. $\dfrac{6}{8} \times \dfrac{1}{2} = \dfrac{6}{16}$

6. $\dfrac{1}{3} \times \dfrac{2}{5} = \dfrac{2}{15}$

7. $\dfrac{9}{12} - \dfrac{8}{12} = \dfrac{1}{12}$

8. $\dfrac{8}{80} + \dfrac{70}{80} = \dfrac{78}{80}$

9. $\dfrac{9}{18} + \dfrac{4}{18} = \dfrac{13}{18}$

$\dfrac{13}{18} + \dfrac{3}{4} = \dfrac{52}{72} + \dfrac{54}{72} = \dfrac{106}{72}$ or $1\dfrac{34}{72}$

10. $73 \times 28 = 2,044$

11. $829 \times 72 = 59,688$
12. $164 \times 53 = 8,692$
13. done
14. $(700) \div (70) = (10)$
15. $(900) \div (20) \approx (40 \text{ or } 45)$
16. $50 \div 2 = 25; \ 25 \times 1 = \$25$
17. $\frac{6}{42} + \frac{7}{42} = \frac{13}{42}$ of his trees
18. $\frac{6}{15} - \frac{5}{15} = \frac{1}{15}$ more
19. $\frac{24}{32} \div \frac{4}{32} = \frac{24 \div 4}{1} = 6$ people
20. $\frac{1}{2} \times \frac{3}{4} = \frac{3}{8}$ cup

Systematic Review 10F

1. $\frac{5}{8} \div \frac{6}{8} = \frac{5 \div 6}{1} = \frac{5}{6}$
2. $\frac{8}{12} \div \frac{3}{12} = \frac{8 \div 3}{1} = \frac{8}{3}$ or $2\frac{2}{3}$
3. $\frac{12}{24} \div \frac{4}{24} = \frac{12 \div 4}{1} = 3$
4. $\frac{4}{5} \times \frac{1}{10} = \frac{4}{50}$
5. $\frac{5}{6} \times \frac{1}{12} = \frac{5}{72}$
6. $\frac{1}{2} \times \frac{1}{4} = \frac{1}{8}$
7. $\frac{5}{20} + \frac{8}{20} = \frac{13}{20}$
8. $\frac{40}{50} - \frac{35}{50} = \frac{5}{50}$
9. $\frac{12}{22} + \frac{11}{22} = \frac{23}{22}$

$\frac{23}{22} + \frac{2}{3} = \frac{69}{66} + \frac{44}{66} = \frac{113}{66}$ or $1\frac{47}{66}$
10. $35 \times 16 = 560$
11. $182 \times 68 = 12,376$
12. $390 \times 41 = 15,990$
13. $(600) \div (40) = (15)$
14. $(400) \div (80) = (5)$
15. $(400) \div (10) = (40)$
16. $\frac{1}{5} \times \frac{2}{3} = \frac{2}{15}$ of the job
17. $\frac{1}{4} \div \frac{1}{16} = \frac{16}{64} \div \frac{4}{64} = \frac{16 \div 4}{1} = 4$ people
18. $\frac{3}{7} + \frac{1}{4} = \frac{12}{28} + \frac{7}{28} = \frac{19}{28}$ of her time

20. $\frac{5}{5} - \frac{1}{5} = \frac{4}{5}$ only words

$\frac{5}{6} \times \frac{4}{5} = \frac{20}{30}$ of the pages

Systematic Review 11A
1. done
2. 1×10
2×5
1, 2, 5, 10
3. 1×18
2×9
3×6
1, 2, 3, 6, 9, 18
4. yes
5. no
6. no
7. no
8. yes
9. yes
10. yes
11. yes
12. no
13. yes
14. no
15. yes
16. yes
17. yes
18. no
19. done
20. 1, 3, 5, 15
1, 2, 5, 10
GCF = 5
21. 1, 2, 3, 6
1, 2, 3, 6, 9, 18
GCF = 6

Lesson Practice 11B

1. 1×9
   3×3
   1, 3, 9
2. 1×4
   2×2
   1, 2, 4
3. 1×14
   2×7
   1, 2, 7, 14
4. yes
5. yes
6. no
7. yes
8. no
9. no
10. done
11. 1×28
    2×14
    4×7
    1, 2, 4, 7, 14, 28
12. $\underline{1}, \underline{3}, 9$
    $\underline{1}, \underline{3}, 5, 15$
    GCF = 3
13. $\underline{1}, 2, \underline{5}, 10$
    $\underline{1}, \underline{5}, 25$
    GCF = 5
14. $\underline{1}, \underline{2}, \underline{3}, \underline{6}$
    $\underline{1}, \underline{2}, \underline{3}, 4, \underline{6}, 12$
    GCF = 6

Lesson Practice 11C

1. 1×8
   2×4
   1, 2, 4, 8
2. 1×20
   2×10
   4×5
   1, 2, 4, 5, 10, 20
3. 1×22
   2×11
   1, 2, 11, 22
4. no

5. no
6. yes
7. yes
8. no
9. no
10. 1×42
    2×21
    3×14
    6×7
    1, 2, 3, 6, 7, 14, 21, 42
11. 1×34
    2×17
    1, 2, 17, 34
12. 1×50
    2×25
    5×10
    1, 2, 5, 10, 25, 50
13. $\underline{1}, \underline{2}, \underline{4}, \underline{8}$
    $\underline{1}, \underline{2}, 3, \underline{4}, 6, \underline{8}, 12, 16, 24, 48$
    GCF = 8
14. $\underline{1}, 3, \underline{5}, 15$
    $\underline{1}, \underline{5}, 7, 35$
    GCF = 5
15. $\underline{1}, 3, \underline{9}$
    $\underline{1}, 2, \underline{3}, 6, \underline{9}, 18$
    GCF = 9

Systematic Review 11D

1. yes
2. yes
3. no
4. $\underline{1}, \underline{2}, 4, \underline{7}, \underline{14}, 28$
   $\underline{1}, \underline{2}, 3, 6, \underline{7}, \underline{14}, 21, 42$
   GCF = 14
5. $\underline{1}, 2, \underline{3}, 6, \underline{9}, 18$
   $\underline{1}, \underline{3}, \underline{9}, 27, 81$
   GCF = 9
6. $\frac{1}{2} \times \frac{1}{3} = \frac{1}{6}$
7. $\frac{2}{3} \times \frac{4}{5} = \frac{8}{15}$
8. $\frac{3}{6} \times \frac{2}{5} = \frac{6}{30}$

9. $\frac{28}{35} \div \frac{5}{35} = \frac{28 \div 5}{1} = \frac{28}{5}$ or $5\frac{3}{5}$

10. $\frac{15}{24} \div \frac{8}{24} = \frac{15 \div 8}{1} = \frac{15}{8}$ or $1\frac{7}{8}$

11. $\frac{24}{30} \div \frac{5}{30} = \frac{24 \div 5}{1} = \frac{24}{5}$ or $4\frac{4}{5}$

12. done

13. $498 \div 51 = 9\frac{39}{51}$

14. $560 \div 28 = 20$

15. $\frac{1}{2} + \frac{1}{4} = \frac{4}{8} + \frac{2}{8} = \frac{6}{8}$ eaten

    $\frac{8}{8} - \frac{6}{8} = \frac{2}{8}$ left

16. $\frac{6}{30} + \frac{10}{30} = \frac{16}{30}$ of the class

17. $36 \div 3 = 12;$

    $12 \times 1 = 12$ girls

    $36 - 12 = 24$ boys

18. $\frac{4}{5} \times \frac{1}{2} = \frac{4}{10}$ of the total pay

Systematic Review 11E

1. yes

2. no

3. $\underline{1}, \underline{5}, 25$

   $\underline{1}, 2, 3, \underline{5}, 6, 10, 15, 30$

   GCF = 5

4. $\underline{1}, \underline{3}, 5, \underline{9}, 15, 45$

   $\underline{1}, \underline{3}, \underline{9}, 27$

   GCF = 9

5. $\frac{2}{3} \times \frac{7}{8} = \frac{14}{24}$

6. $\frac{1}{2} \times \frac{3}{9} = \frac{3}{18}$

7. $\frac{3}{4} \times \frac{5}{6} = \frac{15}{24}$

8. $\frac{20}{32} \div \frac{8}{32} = \frac{20 \div 8}{1} = \frac{20}{8}$ or $2\frac{4}{8}$

9. $\frac{36}{54} \div \frac{9}{54} = \frac{36 \div 9}{1} = 4$

10. $\frac{4}{8} \div \frac{6}{8} = \frac{4 \div 6}{1} = \frac{4}{6}$

11. $\frac{24}{56} < \frac{28}{56}$

12. $\frac{6}{15} > \frac{5}{15}$

13. $\frac{20}{36} < \frac{27}{36}$

14. $424 \div 53 = 8$

15. $711 \div 74 = 9\frac{45}{74}$

16. $890 \div 22 = 40\frac{10}{22}$

17. $25 \times 52 = 1,300$ people

18. $\frac{2}{8} + \frac{1}{8} + \frac{4}{8} = \frac{7}{8}$ completed

    $\frac{8}{8} - \frac{7}{8} = \frac{1}{8}$ to be finished

19. $\frac{24}{32} \div \frac{4}{32} = \frac{24 \div 4}{1} = 6$ sections

20. $\frac{1}{12} \times \frac{1}{4} = \frac{1}{48}$ of this year

Systematic Review 11F

1. no

2. no

3. $\underline{1}, \underline{2}, 4, 8, 16$

   $\underline{1}, \underline{2}, 17, 34$

   GCF = 2

4. $\underline{1}, \underline{2}, 3, \underline{4}, 6, 12$

   $\underline{1}, \underline{2}, \underline{4}, 5, 8, 10, 20, 40$

   GCF = 4

5. $\frac{5}{9} \times \frac{1}{2} = \frac{5}{18}$

6. $\frac{3}{5} \times \frac{1}{4} = \frac{3}{20}$

7. $\frac{6}{7} \times \frac{2}{3} = \frac{12}{21}$

8. $\frac{3}{27} \div \frac{9}{27} = \frac{3 \div 9}{1} = \frac{3}{9}$

9. $\frac{15}{18} \div \frac{12}{18} = \frac{15 \div 12}{1} = \frac{15}{12}$ or $1\frac{3}{12}$

10. $\frac{16}{32} \div \frac{8}{32} = \frac{16 \div 8}{1} = 2$

11. $\frac{6}{12} < \frac{10}{12}$

12. $\frac{40}{80} = \frac{40}{80}$

13. $\frac{9}{108} < \frac{24}{108}$

14. $559 \div 43 = 13$

15. $414 \div 79 = 5\frac{19}{79}$

16. $392 \div 14 = 28$

17. $14 + 14 + 14 + 14 = 56'$ perimeter

$\frac{1}{7}$ of $56 = 8'$ for door openings

$56' - 8' = 48'$ baseboard

18. yes

19. $\frac{4}{32} + \frac{8}{32} = \frac{12}{32}$ of his money

20. $\frac{1}{8}$ of $48 = 6$

$\frac{1}{4}$ of $48 = 12$

$6 + 12 = 18$

$48 - 18 = \$30$

Lesson Practice 12A

1. done

2. $\frac{2}{4} \div \frac{2}{2} = \frac{1}{2}$

3. $\frac{18}{24} \div \frac{6}{6} = \frac{3}{4}$

4. $\frac{8}{12} \div \frac{4}{4} = \frac{2}{3}$

5. done

6. $\frac{6}{8} \div \frac{2}{2} = \frac{3}{4}$

7. $\frac{4}{8} \div \frac{4}{4} = \frac{1}{2}$

8. done

9. $\underline{1}, \underline{2}, \underline{4}$

$\underline{1}, \underline{2}, 3, \underline{4}, 6, 8, 12, 24$

GCF = 4

$\frac{4}{24} \div \frac{4}{4} = \frac{1}{6}$

10. $\underline{1}, \underline{2}, \underline{3}, \underline{6}$

$\underline{1}, \underline{2}, \underline{3}, \underline{6}, 9, 18$

GCF = 6

$\frac{6}{18} \div \frac{6}{6} = \frac{1}{3}$

11. $\underline{1}, \underline{2}, \underline{3}, \underline{6}, 9, 18$

$\underline{1}, \underline{2}, \underline{3}, 5, \underline{6}, 10, 15, 30$

GCF = 6

$\frac{18}{30} \div \frac{6}{6} = \frac{3}{5}$

Lesson Practice 12B

1. $\frac{9}{12} \div \frac{3}{3} = \frac{3}{4}$

2. $\frac{4}{8} \div \frac{4}{4} = \frac{1}{2}$

3. $\frac{8}{10} \div \frac{2}{2} = \frac{4}{5}$

4. $\frac{25}{30} \div \frac{5}{5} = \frac{5}{6}$

5. $\frac{5}{20} \div \frac{5}{5} = \frac{1}{4}$

6. $\frac{10}{12} \div \frac{2}{2} = \frac{5}{6}$

7. $\frac{12}{18} \div \frac{6}{6} = \frac{2}{3}$

8. $\underline{1}, \underline{3}$

$\underline{1}, \underline{3}, 5, 15$

GCF = 3

$\frac{3}{15} \div \frac{3}{3} = \frac{1}{5}$

9. $\underline{1}, \underline{2}, \underline{4}, 8$

$\underline{1}, \underline{2}, 3, \underline{4}, 6, 12$

GCF = 4

$\frac{8}{12} \div \frac{4}{4} = \frac{2}{3}$

10. $\underline{1}, \underline{3}, 5, 15$

$\underline{1}, 2, \underline{3}, 6, 9, 18$

GCF = 3

$\frac{15}{18} \div \frac{3}{3} = \frac{5}{6}$

11. $\underline{1}, 2, \underline{7}, 14$

$\underline{1}, 3, \underline{7}, 21$

GCF = 7

$\frac{14}{21} \div \frac{7}{7} = \frac{2}{3}$

Lesson Practice 12C

1. $\frac{10}{15} \div \frac{5}{5} = \frac{2}{3}$

2. $\frac{9}{15} \div \frac{3}{3} = \frac{3}{5}$

3. $\frac{27}{33} \div \frac{3}{3} = \frac{9}{11}$

4. $\frac{20}{36} \div \frac{4}{4} = \frac{5}{9}$

5. $\frac{42}{48} \div \frac{6}{6} = \frac{7}{8}$

6. $\underline{1}, 2, \underline{5}, 10$
    $\underline{1}, 3, \underline{5}, 15$
    GCF = 5
    $\frac{10}{15} \div \frac{5}{5} = \frac{2}{3}$

7. $\underline{1}, \underline{2}, \underline{4}, \underline{8}, 16$
    $\underline{1}, \underline{2}, 3, \underline{4}, 6, \underline{8}, 12, 24$
    GCF = 8
    $\frac{16}{24} \div \frac{8}{8} = \frac{2}{3}$

8. $\frac{25}{30} \div \frac{5}{5} = \frac{5}{6}$

9. $\frac{6}{8} \div \frac{2}{2} = \frac{3}{4}$

10. $\frac{10}{16} \div \frac{2}{2} = \frac{5}{8}$

11. $\frac{18}{21} \div \frac{3}{3} = \frac{6}{7}$

12. $\frac{45}{50} \div \frac{5}{5} = \frac{9}{10}$

13. $\frac{27}{36} \div \frac{9}{9} = \frac{3}{4}$

14. $\frac{7}{14} \div \frac{7}{7} = \frac{1}{2}$
    $\frac{1}{2}$ of 56 = 28 people

Systematic Review 12D

1. $\frac{12}{28} \div \frac{4}{4} = \frac{3}{7}$

2. $\frac{14}{49} \div \frac{7}{7} = \frac{2}{7}$

3. $\frac{35}{50} \div \frac{5}{5} = \frac{7}{10}$

4. yes

5. yes

6. $\frac{24}{32} + \frac{4}{32} = \frac{28}{32} = \frac{7}{8}$

7. $\frac{5}{35} + \frac{14}{35} = \frac{19}{35}$

8. $\frac{6}{48} \div \frac{6}{6} = \frac{1}{8}$

9. $\frac{4}{50} \div \frac{2}{2} = \frac{2}{25}$

10. $\frac{8}{32} \div \frac{20}{32} = \frac{8 \div 20}{1} = \frac{8}{20} = \frac{2}{5}$

11. $\frac{6}{15} \div \frac{10}{15} = \frac{6 \div 10}{1} = \frac{6}{10} = \frac{3}{5}$

12. $531 \times 624 = 331,344$

13. $3,728 \times 128 = 477,184$

14. $6,593 \times 756 = 4,984,308$

15. $\frac{2}{4} \div \frac{2}{2} = \frac{1}{2}$ of an orange

16. $\frac{4}{6} \div \frac{1}{6} = \frac{4 \div 1}{1} = 4$ helpers

17. $\frac{3}{5} \times \frac{2}{3} = \frac{6}{15} = \frac{2}{5}$ of the letters

18. $\frac{2}{3} + \frac{1}{5} = \frac{10}{15} + \frac{3}{15} = \frac{13}{15}$ are correct

Systematic Review 12E

1. $\frac{24}{30} \div \frac{6}{6} = \frac{4}{5}$

2. $\frac{18}{28} \div \frac{2}{2} = \frac{9}{14}$

3. $\frac{15}{35} \div \frac{5}{5} = \frac{3}{7}$

4. no

5. no

6. $\frac{8}{16} + \frac{6}{16} = \frac{14}{16} = \frac{7}{8}$

7. $\frac{20}{50} + \frac{15}{50} = \frac{35}{50} = \frac{7}{10}$

8. $\frac{10}{24} = \frac{5}{12}$

9. $\frac{18}{32} = \frac{9}{16}$

10. $\frac{9}{54} \div \frac{30}{54} = \frac{9 \div 30}{1} = \frac{9}{30} = \frac{3}{10}$

11. $\frac{4}{8} \div \frac{6}{8} = \frac{4 \div 6}{1} = \frac{4}{6} = \frac{2}{3}$

12. $728 \times 165 = 120,120$

13. $2,192 \times 864 = 1,893,888$

14. $8,651 \times 549 = 4,749,399$

15. $\frac{5}{15} \div \frac{5}{5} = \frac{1}{3}$

16. $512 \times 150 = 76,800$ paper clips

17. $\frac{2}{3} + \frac{1}{5} = \frac{10}{15} + \frac{3}{15} = \frac{13}{15}$ of the beads

18. $360 \div 30 = 12$ paychecks

19. $\frac{3}{4} \times \frac{1}{2} = \frac{3}{8}$ yd

20. $\frac{3}{4} - \frac{3}{8} = \frac{24}{32} - \frac{12}{32} = \frac{12}{32} = \frac{3}{8}$ yd

Systematic Review 12F

1. $\dfrac{18}{63} \div \dfrac{9}{9} = \dfrac{2}{7}$

2. $\dfrac{26}{32} \div \dfrac{2}{2} = \dfrac{13}{16}$

3. $\dfrac{42}{48} \div \dfrac{6}{6} = \dfrac{7}{8}$

4. yes

5. yes

6. $\dfrac{36}{72} + \dfrac{8}{72} = \dfrac{44}{72} = \dfrac{11}{18}$

7. $\dfrac{20}{30} + \dfrac{6}{30} = \dfrac{26}{30} = \dfrac{13}{15}$

8. $\dfrac{12}{42} \div \dfrac{6}{6} = \dfrac{2}{7}$

9. $\dfrac{4}{20} \div \dfrac{4}{4} = \dfrac{1}{5}$

10. $\dfrac{8}{32} \div \dfrac{28}{32} = \dfrac{8 \div 28}{1} = \dfrac{8}{28} = \dfrac{2}{7}$

11. $\dfrac{3}{6} \div \dfrac{4}{6} = \dfrac{3 \div 4}{1} = \dfrac{3}{4}$

12. $371 \times 244 = 90,524$

13. $5,970 \times 186 = 1,110,420$

14. $7,035 \times 369 = 2,595,915$

15. $\dfrac{18}{30} \div \dfrac{6}{6} = \dfrac{3}{5}$

    $500 \div 5 = 100;\ 100 \times 3 = 300$ people

16. $\dfrac{16}{60} > \dfrac{15}{60}$; more sopranos

17. $16 \times 12 = 192;\ 192 \div 3 = 64$ packs

18. $\dfrac{3}{4}$ of $64 = \$48$

19. $\dfrac{12}{18} \div \dfrac{6}{18} = \dfrac{12 \div 6}{1} = 2$ pieces

20. $\dfrac{1}{3} \div \dfrac{4}{1} = \dfrac{1}{3} \div \dfrac{12}{3} = \dfrac{1}{12}$ of a pie

Lesson Practice 13A

1. done

2. $3 \times 3 \times 5$

3. $2 \times 3 \times 5$

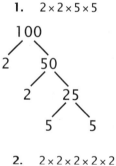

4. done

5. $3 \times 3 \times 3$

6. $2 \times 5 \times 5$

7. done

8. $\dfrac{20}{28} = \dfrac{2 \times 2 \times 5}{2 \times 2 \times 7} = \dfrac{2}{2} \times \dfrac{2}{2} \times \dfrac{5}{7} = \dfrac{5}{7}$

9. $\dfrac{30}{50} = \dfrac{2 \times 3 \times 5}{2 \times 5 \times 5} = \dfrac{2}{2} \times \dfrac{5}{5} \times \dfrac{3}{5} = \dfrac{3}{5}$

10. $\dfrac{21}{33} = \dfrac{3 \times 7}{3 \times 11} = \dfrac{3}{3} \times \dfrac{7}{11} = \dfrac{7}{11}$

11. $\dfrac{18}{26} = \dfrac{2 \times 3 \times 3}{2 \times 13} = \dfrac{2}{2} \times \dfrac{3 \times 3}{13} = \dfrac{9}{13}$

12. $\dfrac{24}{54} = \dfrac{2 \times 2 \times 2 \times 3}{2 \times 3 \times 3 \times 3} = \dfrac{2}{2} \times \dfrac{3}{3} \times$

    $\dfrac{2 \times 2}{3 \times 3} = \dfrac{4}{9}$

Lesson Practice 13B

1. $2 \times 2 \times 5 \times 5$

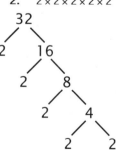

2. $2 \times 2 \times 2 \times 2 \times 2$

3. $2 \times 2 \times 2 \times 3 \times 3$

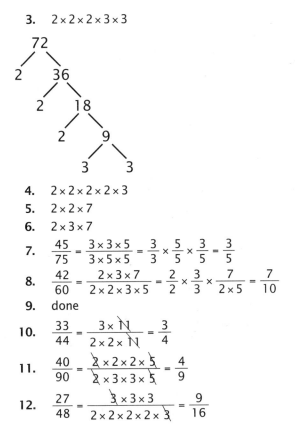

4. $2 \times 2 \times 2 \times 2 \times 3$

5. $2 \times 2 \times 7$

6. $2 \times 3 \times 7$

7. $\dfrac{45}{75} = \dfrac{3 \times 3 \times 5}{3 \times 5 \times 5} = \dfrac{3}{3} \times \dfrac{5}{5} \times \dfrac{3}{5} = \dfrac{3}{5}$

8. $\dfrac{42}{60} = \dfrac{2 \times 3 \times 7}{2 \times 2 \times 3 \times 5} = \dfrac{2}{2} \times \dfrac{3}{3} \times \dfrac{7}{2 \times 5} = \dfrac{7}{10}$

9. done

10. $\dfrac{33}{44} = \dfrac{3 \times \cancel{11}}{2 \times 2 \times \cancel{11}} = \dfrac{3}{4}$

11. $\dfrac{40}{90} = \dfrac{\cancel{2} \times 2 \times 2 \times \cancel{5}}{\cancel{2} \times 3 \times 3 \times \cancel{5}} = \dfrac{4}{9}$

12. $\dfrac{27}{48} = \dfrac{\cancel{3} \times 3 \times 3}{2 \times 2 \times 2 \times 2 \times \cancel{3}} = \dfrac{9}{16}$

Lesson Practice 13C

1. $2 \times 2 \times 5$

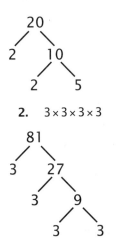

2. $3 \times 3 \times 3 \times 3$

3. $2 \times 2 \times 13$

(tree: 52 → 2, 26 → 2, 13)

4. $2 \times 3 \times 11$

5. $2 \times 2 \times 11$

6. $2 \times 2 \times 2 \times 2 \times 5$

7. $\dfrac{20}{24} = \dfrac{\cancel{2} \times \cancel{2} \times 5}{\cancel{2} \times \cancel{2} \times 2 \times 3} = \dfrac{5}{6}$

8. $\dfrac{30}{36} = \dfrac{\cancel{2} \times \cancel{3} \times 5}{\cancel{2} \times 2 \times \cancel{3} \times 3} = \dfrac{5}{6}$

9. $\dfrac{48}{54} = \dfrac{2 \times 2 \times 2 \times \cancel{2} \times \cancel{3}}{\cancel{2} \times \cancel{3} \times 3 \times 3} = \dfrac{8}{9}$

10. $\dfrac{9}{27} = \dfrac{\cancel{3} \times \cancel{3}}{3 \times \cancel{3} \times \cancel{3}} = \dfrac{1}{3}$

11. $\dfrac{15}{25} = \dfrac{3 \times \cancel{5}}{5 \times \cancel{5}} = \dfrac{3}{5}$

12. $\dfrac{12}{15} = \dfrac{2 \times 2 \times \cancel{3}}{\cancel{3} \times 5} = \dfrac{4}{5}$

Systematic Review 13D

1. 2×13

2. $2 \times 2 \times 3 \times 5$

3. $2 \times 2 \times 2 \times 3$

4. $\dfrac{18}{24} = \dfrac{\cancel{2} \times \cancel{3} \times 3}{\cancel{2} \times 2 \times 2 \times \cancel{3}} = \dfrac{3}{4}$

5. $\dfrac{15}{30} = \dfrac{\cancel{3} \times \cancel{5}}{2 \times \cancel{3} \times \cancel{5}} = \dfrac{1}{2}$

6. $\dfrac{8}{32} + \dfrac{8}{32} = \dfrac{16}{32} \div \dfrac{16}{16} = \dfrac{1}{2}$

7. $\dfrac{10}{16} - \dfrac{8}{16} = \dfrac{2}{16} \div \dfrac{2}{2} = \dfrac{1}{8}$

8. $\dfrac{6}{21} + \dfrac{7}{21} = \dfrac{13}{21}$

9. $\dfrac{2}{12} \div \dfrac{6}{12} = \dfrac{2 \div 6}{1} = \dfrac{2}{6} \div \dfrac{2}{2} = \dfrac{1}{3}$

10. $\dfrac{5}{9} \times \dfrac{3}{7} = \dfrac{15}{63} \div \dfrac{3}{3} = \dfrac{5}{21}$

11. $\dfrac{8}{11} \times \dfrac{3}{4} = \dfrac{24}{44} \div \dfrac{4}{4} = \dfrac{6}{11}$

12. done

13. $9442 \div 53 = 178\dfrac{8}{53}$

14. $4925 \div 189 = 26\frac{11}{189}$

15. $\frac{1}{6} \times 12 = 2$ months

16. $\frac{3}{4} + \frac{1}{8} = \frac{24}{32} + \frac{4}{32} = \frac{28}{32} \div \frac{4}{4} = \frac{7}{8}$ yd

17. $2{,}845 \div 5 = 569$ gal

18. $528 \times 73 = 38{,}544$ mosquitoes

Systematic Review 13E

1. $3 \times 3 \times 3 \times 3$

2. $2 \times 3 \times 3 \times 5$

3. $3 \times 5 \times 5$

4. $\frac{8}{22} = \frac{2 \times 2 \times 2}{2 \times 11} = \frac{4}{11}$

5. $\frac{32}{48} = \frac{2 \times 2 \times 2 \times 2 \times 2}{2 \times 2 \times 2 \times 2 \times 3} = \frac{2}{3}$

6. $\frac{6}{12} + \frac{4}{12} = \frac{10}{12} \div \frac{2}{2} = \frac{5}{6}$

7. $\frac{18}{27} - \frac{9}{27} = \frac{9}{27} \div \frac{9}{9} = \frac{1}{3}$

8. $\frac{30}{50} + \frac{5}{50} = \frac{35}{50} \div \frac{5}{5} = \frac{7}{10}$

9. $\frac{16}{64} \div \frac{48}{64} = \frac{16 \div 48}{1} = \frac{16}{48} \div \frac{16}{16} = \frac{1}{3}$

10. $\frac{5}{9} \times \frac{2}{6} = \frac{10}{54} \div \frac{2}{2} = \frac{5}{27}$

11. $\frac{2}{5} \times \frac{3}{4} = \frac{6}{20} \div \frac{2}{2} = \frac{3}{10}$

12. $\frac{30}{48} < \frac{32}{48}$

13. $\frac{7}{21} > \frac{6}{21}$

14. $\frac{40}{80} = \frac{40}{80}$

15. $1130 \div 38 = 29\frac{28}{38}$ or $29\frac{14}{19}$

16. $2686 \div 22 = 122\frac{2}{22}$ or $122\frac{1}{11}$

17. $5032 \div 235 = 21\frac{97}{235}$

18. $45 \times 365 = 16{,}425$ pennies

19. $\frac{12}{36} + \frac{15}{36} = \frac{27}{36}$ of an hour

20. $\frac{27}{36}$ of 60 : $\frac{27}{36} \times \frac{60}{1} = \frac{1{,}620}{36} = 45$ min

$\frac{3}{4}$ of 60 : $60 \div 4 = 15$; $15 \times 3 = 45$ min

The simplified fraction is easier to use.

Systematic Review 13F

1. $2 \times 2 \times 2 \times 2 \times 2 \times 2$

2. $2 \times 2 \times 2 \times 2$

3. $3 \times 3 \times 5$

4. $\frac{81}{90} = \frac{3 \times 3 \times 3 \times 3}{2 \times 3 \times 3 \times 5} = \frac{9}{10}$

5. $\frac{12}{18} = \frac{2 \times 2 \times 3}{2 \times 3 \times 3} = \frac{2}{3}$

6. $\frac{18}{48} + \frac{8}{48} = \frac{26}{48} \div \frac{2}{2} = \frac{13}{24}$

7. $\frac{45}{50} - \frac{20}{50} = \frac{25}{50} \div \frac{25}{25} = \frac{1}{2}$

8. $\frac{12}{24} + \frac{6}{24} = \frac{18}{24} \div \frac{6}{6} = \frac{3}{4}$

9. $\frac{10}{50} \div \frac{20}{50} = \frac{10 \div 20}{1} = \frac{10}{20} \div \frac{10}{10} = \frac{1}{2}$

10. $\frac{4}{7} \times \frac{2}{8} = \frac{8}{56} \div \frac{8}{8} = \frac{1}{7}$

11. $\frac{5}{12} \times \frac{3}{5} = \frac{15}{60} \div \frac{15}{15} = \frac{1}{4}$

12. $\frac{60}{110} < \frac{77}{110}$

13. $\frac{9}{36} < \frac{16}{36}$

14. $\frac{15}{36} > \frac{12}{36}$

15. $3621 \div 96 = 37\frac{69}{96}$ or $37\frac{23}{32}$

16. $7192 \div 73 = 98\frac{38}{73}$

17. $6831 \div 120 = 56\frac{111}{120}$ or $56\frac{37}{40}$

18. $2{,}075 \div 25 = 83$ days

19. $\frac{6}{18} + \frac{3}{18} = \frac{9}{18} = \frac{1}{2}$

$\frac{1}{2} + \frac{1}{8} = \frac{8}{16} + \frac{2}{16} = \frac{10}{16} = \frac{5}{8}$ of the job

20. $\frac{8}{8} - \frac{5}{8} = \frac{3}{8}$ of the job left

Lesson Practice 14A

1. done
2. $\frac{7}{10}$
3. $\frac{2}{3}$
4. $\frac{3}{7}$
5. done
6.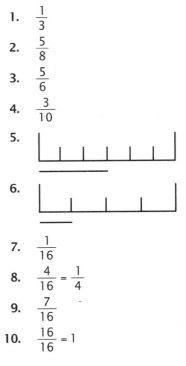
7. done
8. $\frac{12}{16} = \frac{3}{4}$
9. $\frac{8}{16} = \frac{1}{2}$
10. $\frac{3}{16}$
11. $\frac{14}{16} = \frac{7}{8}$
12. $\frac{10}{16} = \frac{5}{8}$

Lesson Practice 14B

1. $\frac{1}{3}$
2. $\frac{5}{8}$
3. $\frac{5}{6}$
4. $\frac{3}{10}$
5.
6.
7. $\frac{1}{16}$
8. $\frac{4}{16} = \frac{1}{4}$
9. $\frac{7}{16}$
10. $\frac{16}{16} = 1$

11. $\frac{15}{16}$
12. $\frac{8}{16} = \frac{1}{2}$

Lesson Practice 14C

1. $\frac{6}{10}$
2. $\frac{4}{7}$
3. $\frac{3}{6}$
4. $\frac{1}{4}$
5.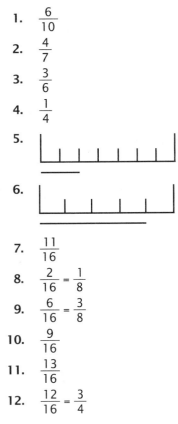
6.
7. $\frac{11}{16}$
8. $\frac{2}{16} = \frac{1}{8}$
9. $\frac{6}{16} = \frac{3}{8}$
10. $\frac{9}{16}$
11. $\frac{13}{16}$
12. $\frac{12}{16} = \frac{3}{4}$

Systematic Review 14D

1. $\frac{7}{8}$
2. $\frac{3}{6} = \frac{1}{2}$
3. $\frac{5}{16}$
4. $\frac{14}{16} = \frac{7}{8}$
5. $2 \times 2 \times 7$
6. 5×11
7. $2 \times 2 \times 3 \times 7$

8. $\dfrac{48}{64} = \dfrac{\cancel{2} \times \cancel{2} \times \cancel{2} \times \cancel{2} \times 3}{\cancel{2} \times \cancel{2} \times \cancel{2} \times \cancel{2} \times 2 \times 2} = \dfrac{3}{4}$

9. done

10. $20 \times 10 = 200$ sq ft

11. $15 \times 7 = 105$ sq yd

12. no

13. yes

14. $12 : \underline{2}, \underline{3}, 4, \underline{6}, 12$
    $18 : \underline{2}, \underline{3}, \underline{6}, 9, 18$
    GCF = 6

15. $25 : \underline{5}, \underline{25}$
    $50 : 2, \underline{5}, \underline{25}, 50$
    GCF = 25

16. $15 \times 13 = 195$ sq ft

Systematic Review 14E

1. $\dfrac{2}{6} = \dfrac{1}{3}$

2. $\dfrac{6}{7}$

3. $\dfrac{10}{16} = \dfrac{5}{8}$

4. $\dfrac{3}{16}$

5. $2 \times 5 \times 7$

6. $3 \times 3 \times 5$

7. $2 \times 3 \times 5$

8. $\dfrac{33}{63} = \dfrac{\cancel{3} \times 11}{\cancel{3} \times 3 \times 7} = \dfrac{11}{21}$

9. $8 \times 5 = 40$ sq in

10. $90 \times 45 = 4,050$ sq ft

11. $28 \times 15 = 420$ sq yd

12. no

13. $21 : \underline{3}, \underline{7}, \underline{21}$
    $42 : 2, \underline{3}, 6, \underline{7}, 14, \underline{21}, 42$
    GCF = 21

14. $\dfrac{3}{4} \times \dfrac{1}{3} = \dfrac{3}{12} \div \dfrac{3}{3} = \dfrac{1}{4}$ for school clothes

15. $\dfrac{8}{12} < \dfrac{9}{12}$ so $\dfrac{2}{3} < \dfrac{3}{4}$

16. $\dfrac{9}{12} - \dfrac{8}{12} = \dfrac{1}{12}$ of a cup

17. $15 + 13 + 15 + 13 = 56$ ft

18. $56 \div 4 = 14$
    $14 \times 3 = 42$
    $56 - 42 = 14$ ft needed

Systematic Review 14F

1. $\dfrac{2}{3}$

2. $\dfrac{4}{10} = \dfrac{2}{5}$

3. $\dfrac{7}{16}$

4. $\dfrac{16}{16} = 1$

5. $2 \times 3 \times 11$

6. $2 \times 2 \times 7$

7. $2 \times 3 \times 3 \times 3$

8. $\dfrac{75}{100} = \dfrac{3 \times \cancel{5} \times \cancel{5}}{2 \times 2 \times \cancel{5} \times \cancel{5}} = \dfrac{3}{4}$

9. $3 \times 2 = 6$ sq in

10. $11 \times 6 = 66$ sq ft

11. $35 \times 19 = 665$ sq yd

12. yes

13. $24 : \underline{2}, \underline{3}, \underline{4}, \underline{6}, 8, \underline{12}, 24$
    $36 : \underline{2}, \underline{3}, \underline{4}, \underline{6}, 9, \underline{12}, 18, 36$
    GCF = 12

14. $\dfrac{18}{42} + \dfrac{7}{42} = \dfrac{25}{42}$ of the chocolates; no

15. $\dfrac{8}{12} \div \dfrac{1}{6} = \dfrac{48}{72} \div \dfrac{12}{72} = \dfrac{48 \div 12}{1} = 4$ times

16. $36 \times 12 = 432$ eggs

17. $2,160 \div 432 = 5$ crates

18. area

Lesson Practice 15A

1. done

2. $\dfrac{6}{6} + \dfrac{5}{6} = \dfrac{11}{6}$

3. $\dfrac{2}{2} + \dfrac{2}{2} + \dfrac{1}{2} = \dfrac{5}{2}$

4. done

5. $1 + 1 + \dfrac{1}{3} = 2\dfrac{1}{3}$

6. $1 + \dfrac{4}{5} = 1\dfrac{4}{5}$

Lesson Practice 15B

1. $\frac{4}{4} + \frac{1}{4} = \frac{5}{4}$

2. $\frac{6}{6} + \frac{6}{6} + \frac{2}{6} = \frac{14}{6}$

3. done

4. $\frac{3}{3} + \frac{3}{3} + \frac{3}{3} + \frac{2}{3} = \frac{11}{3}$

5. $1 + 1 + \frac{3}{5} = 2\frac{3}{5}$

6. done

7. $\frac{2}{2} + \frac{1}{2} = 1\frac{1}{2}$

8. $\frac{6}{6} + \frac{6}{6} + \frac{6}{6} + \frac{5}{6} =$
   $1 + 1 + 1 + \frac{5}{6} = 3\frac{5}{6}$

Lesson Practice 15C

1. $\frac{6}{6} + \frac{6}{6} + \frac{3}{6} = \frac{15}{6}$

2. $\frac{2}{2} + \frac{2}{2} + \frac{1}{2} = \frac{5}{2}$

3. $\frac{5}{5} + \frac{4}{5} = \frac{9}{5}$

4. $\frac{8}{8} + \frac{8}{8} + \frac{8}{8} + \frac{5}{8} = \frac{29}{8}$

5. $\frac{4}{4} + \frac{4}{4} + \frac{1}{4} = \frac{9}{4}$

6. $1 + 1 + \frac{2}{3} = 2\frac{2}{3}$

7. done

8. $\frac{7}{7} + \frac{1}{7} = 1\frac{1}{7}$

9. $\frac{2}{2} + \frac{2}{2} + \frac{1}{2} = 2\frac{1}{2}$

10. $\frac{4}{4} + \frac{4}{4} + \frac{4}{4} + \frac{4}{4} + \frac{1}{4} = 4\frac{1}{4}$

Systematic Review 15D

1. $\frac{3}{3} + \frac{3}{3} + \frac{2}{3} = \frac{8}{3}$

2. $\frac{2}{2} + \frac{2}{2} + \frac{2}{2} + \frac{1}{2} = \frac{7}{2}$

3. $\frac{9}{9} + \frac{4}{9} = 1\frac{4}{9}$

4. $\frac{5}{5} + \frac{5}{5} + \frac{1}{5} = 2\frac{1}{5}$

5. $\frac{15}{16}$

6. $\frac{4}{16} = \frac{1}{4}$

7. $\frac{16}{24} \div \frac{8}{8} = \frac{2}{3}$

8. $\frac{28}{35} \div \frac{7}{7} = \frac{4}{5}$

9. $\frac{54}{63} \div \frac{9}{9} = \frac{6}{7}$

10. done

11. done

12. $10 \times 10 = 100$ sq in

13. $10 + 10 + 10 + 10 = 40$ in

14. $5 \times 5 = 25$ sq mi

15. $5 + 5 + 5 + 5 = 20$ mi

16. 3×31

17. 44: 2, 4, 11, 22, 44
    55: 5, 11, 55
    GCF = 11

18. $\frac{2}{16} = \frac{1}{8}$
    $\frac{7}{8} \div \frac{1}{8} = \frac{7 \div 1}{1} = 7$ sections

Systematic Review 15E

1. $\frac{6}{6} + \frac{5}{6} = \frac{11}{6}$

2. $\frac{9}{9} + \frac{9}{9} + \frac{9}{9} + \frac{4}{9} = \frac{31}{9}$

3. $\frac{8}{8} + \frac{7}{8} = 1\frac{7}{8}$

4. $\frac{10}{10} + \frac{10}{10} + \frac{10}{10} + \frac{3}{10} = 3\frac{3}{10}$

5. $\frac{9}{16}$

6. $\frac{12}{16} = \frac{3}{4}$

7. $\frac{27}{45} \div \frac{9}{9} = \frac{3}{5}$

8. $\frac{44}{60} \div \frac{4}{4} = \frac{11}{15}$

9. $\frac{24}{30} \div \frac{6}{6} = \frac{4}{5}$

10. done

11. $11^2 = 11 \times 11 = 121$ sq mi

12. $14^2 = 14 \times 14 = 196$ sq in

13. $3 \times 3 \times 3 \times 3$

14. $18 : \underline{2}, 3, 6, 9, 18$

$28 : \underline{2}, 4, 7, 14, 28$

GCF = 2

15. no

16. $\frac{1}{3} \times \frac{5}{6} = \frac{5}{18}$ of the people

17. $\$35 \times 175 = \$6,125$

18. $12 + 12 + 12 + 12 = 48$ mi

Systematic Review 15F

1. $\frac{8}{8} + \frac{3}{8} = \frac{11}{8}$

2. $\frac{7}{7} + \frac{7}{7} + \frac{7}{7} + \frac{2}{7} = \frac{23}{7}$

3. $\frac{7}{7} + \frac{4}{7} = 1\frac{4}{7}$

4. $\frac{6}{6} + \frac{6}{6} + \frac{6}{6} + \frac{6}{6} + \frac{4}{6} = 4\frac{4}{6}$

5. $\frac{6}{16} = \frac{3}{8}$

6. $\frac{5}{16}$

7. $\frac{50}{70} \div \frac{10}{10} = \frac{5}{7}$

8. $\frac{24}{36} \div \frac{12}{12} = \frac{2}{3}$

9. $\frac{3}{9} \div \frac{3}{3} = \frac{1}{3}$

10. $25^2 = 625$ sq ft

11. $2^2 = 4$ sq in

12. $17^2 = 289$ sq ft

13. done

14. $3^2 = 9$

15. $10^2 = 100$

16. $21 : \underline{3}, 7, 21$

$24 : 2, \underline{3}, 4, 6, 8, 12, 24$

GCF = $\underline{3}$

17. yes

18. $2 \times 5 \times 5$

19. $\frac{3}{6} + \frac{4}{6} = \frac{7}{6} = 1\frac{1}{6}$ pies

20. $8,925 \div 425 = 21$ days

Lesson Practice 16A

1. done

2. $4\frac{3}{10}$

3. $2\frac{1}{8}$

4. $1\frac{5}{8}$

5. $\frac{3}{8}$

6. $4\frac{3}{16}$

Lesson Practice 16B

1. $1\frac{4}{5}$

2. $3\frac{1}{5}$

3. $4\frac{3}{4}$

4. $2\frac{7}{16}$

5. $1\frac{1}{2}$

6. $3\frac{1}{4}$

Lesson Practice 16C

1. $2\frac{4}{10} = 2\frac{2}{5}$

2. $1\frac{6}{10} = 1\frac{3}{5}$

3. $\frac{1}{2}$

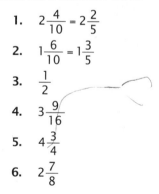

4. $3\frac{9}{16}$

5. $4\frac{3}{4}$

6. $2\frac{7}{8}$

Systematic Review 16D

1. $3\frac{9}{10}$

2. $\frac{8}{8} + \frac{8}{8} + \frac{6}{8} = \frac{22}{8}$

3. $\frac{6}{6} + \frac{6}{6} + \frac{6}{6} + \frac{1}{6} = \frac{19}{6}$

4. $\frac{5}{5} + \frac{2}{5} = 1\frac{2}{5}$

5. $\frac{8}{8} + \frac{8}{8} + \frac{3}{8} = 2\frac{3}{8}$

6. $\frac{15}{30} \div \frac{15}{15} = \frac{1}{2}$

7. $\frac{28}{70} \div \frac{14}{14} = \frac{2}{5}$

8. $\frac{6}{9} \div \frac{3}{3} = \frac{2}{3}$

9. $\frac{15}{40} + \frac{16}{40} = \frac{31}{40}$

10. $\frac{10}{14} - \frac{7}{14} = \frac{3}{14}$

11. $\frac{4}{8} + \frac{6}{8} + \frac{7}{8} = \frac{17}{8} = 2\frac{1}{8}$

12. done

13. $\left(\frac{1}{2}\right)(9)(12) = 54$ sq in

14. $\left(\frac{1}{2}\right)(30)(7) = 105$ sq ft

15. $3 \times 5 \times 5$

16. $15^2 = 225$ sq. ft

17. $\frac{1}{2} \times \frac{4}{10} = \frac{4}{20} = \frac{1}{5}$ of the house

18. $3,279 \div 7 = 468\frac{3}{7}$ weeks

Systematic Review 16E

1. $4\frac{1}{4}$

2. $\frac{10}{10} + \frac{10}{10} + \frac{1}{10} = \frac{21}{10}$

3. $\frac{3}{3} + \frac{3}{3} + \frac{3}{3} + \frac{2}{3} = \frac{11}{3}$

4. $\frac{9}{9} + \frac{8}{9} = 1\frac{8}{9}$

5. $\frac{2}{2} + \frac{2}{2} + \frac{1}{2} = 2\frac{1}{2}$

6. $\frac{20}{32} \div \frac{8}{32} = \frac{20 \div 8}{1} = \frac{20}{8} = \frac{5}{2} = 2\frac{1}{2}$

7. $\frac{10}{14} \div \frac{7}{14} = \frac{10 \div 7}{1} = \frac{10}{7} = 1\frac{3}{7}$

8. $\frac{6}{8} \div \frac{4}{8} = \frac{6 \div 4}{1} = \frac{6}{4} = \frac{3}{2} = 1\frac{1}{2}$

9. $\frac{1}{5} = \frac{2}{10} = \frac{3}{15}$

10. $\frac{3}{4} = \frac{6}{8} = \frac{9}{12}$

11. $\frac{4}{7} = \frac{8}{14} = \frac{12}{21}$

12. $\left(\frac{1}{2}\right)(6)(4) = 12$ sq ft

13. $\left(\frac{1}{2}\right)(8)(6) = 24$ sq in

14. $\frac{1}{2}(21)(5) = 52\frac{1}{2}$ sq ft

15. $15 : 3, \underline{5}, 15$
    $20 : 2, 4, \underline{5}, 10, 20$
    GCF = 5

16. $15 \times 15 = 225$

17. $\frac{1}{2} + \frac{1}{2} + \frac{1}{2} = \frac{3}{2} = 1\frac{1}{2}$ in

18. $256 \times 38 = 9,728$ sq yd

19. $\frac{3}{15} \div \frac{3}{3} = \frac{1}{5}$ of her income

20. $\frac{60}{72} < \frac{66}{72}$;
    Danny has mowed more of the lawn.

Systematic Review 16F

1. $2\frac{3}{8}$

2. $\frac{4}{4} + \frac{4}{4} + \frac{3}{4} = \frac{11}{4}$

3. $\frac{7}{7} + \frac{7}{7} + \frac{7}{7} + \frac{5}{7} = \frac{26}{7}$

4. $\frac{6}{6} + \frac{5}{6} = 1\frac{5}{6}$

5. $\frac{10}{10} + \frac{10}{10} + \frac{7}{10} = 2\frac{7}{10}$

6. $\frac{1}{5} \times \frac{5}{6} = \frac{5}{30} \div \frac{5}{5} = \frac{1}{6}$

7. $\frac{2}{3} \times \frac{4}{5} = \frac{8}{15}$

8. $\frac{1}{3} \times \frac{2}{8} = \frac{2}{24} \div \frac{2}{2} = \frac{1}{12}$

9. $\frac{8}{12} < \frac{9}{12}$

10. $\frac{21}{56} > \frac{8}{56}$

11. $\frac{55}{99} > \frac{36}{99}$

12. $9 \times 3 = 27$ sq ft

13. $9 + 3 + 9 + 3 = 24$ ft

14. $17^2 = 289$ sq. in

15. $17 + 17 + 17 + 17 = 68$ in

16. $\left(\frac{1}{2}\right)(16)(12) = 96$ sq ft

17. $12 + 16 + 20 = 48$ ft

18. $2 \times 2 \times 5 \times 5$

19. $20 \times 20 = 400$

20. $\frac{80}{128} \div \frac{8}{128} = \frac{80 \div 8}{1} = 10$ pieces

Lesson Practice 17A

1. done

2. $(3) + (3) = (6); 5\frac{5}{6}$

3. $(7) + (9) = (16); 16\frac{4}{5}$

4. $(2) + (2) = (4); 3\frac{8}{9}$

5. $(4) + (9) = (13); 12\frac{6}{7}$

6. $(7) + (7) = (14); 13\frac{3}{4}$

7. done

8. $(7) - (3) = (4); 3\frac{3}{8}$

9. $(11) - (10) = (1); 1\frac{1}{5}$

10. $(9) - (5) = (4); 4\frac{2}{9}$

11. $(6) - (4) = (2); 2\frac{1}{4}$

12. $(4) - (3) = (1); 1\frac{1}{6}$

Lesson Practice 17B

1. $(10) + (9) = (19); 18\frac{5}{6}$

2. $(8) + (4) = (12); 11\frac{7}{8}$

3. $(7) + (4) = (11); 11\frac{2}{3}$

4. $(6) + (4) = (10); 9\frac{3}{4}$

5. $(2) + (5) = (7); 6\frac{4}{5}$

6. $(10) + (4) = (14); 13\frac{6}{7}$

7. $(3) - (1) = (2); 1\frac{2}{5}$

8. $(9) - (6) = (3); 2\frac{3}{8}$

9. $(6) - (4) = (2); 1\frac{1}{5}$

10. $(5) - (3) = (2); 1\frac{4}{7}$

11. $(10) - (8) = (2); 2\frac{1}{6}$

12. $(17) - (8) = (9); 8\frac{4}{9}$

13. $2\frac{1}{3} + 1\frac{1}{3} = 3\frac{2}{3}$ pies

14. $4\frac{3}{5} - 3\frac{1}{5} = 1\frac{2}{5}$ lb

Lesson Practice 17C

1. $(5) + (8) = (13); 12\frac{7}{8}$

2. $(4) + (4) = (8); 7\frac{8}{9}$

3. $(8) + (6) = (14); 13\frac{3}{4}$

4. $(2) + (6) = (8); 8\frac{7}{10}$

5. $(9) + (1) = (10); 10\frac{3}{5}$

6. $(6) + (3) = (9); 9\frac{4}{9}$

7. $(13) - (4) = (9); 8\frac{2}{7}$

8. $(8) - (5) = (3); 3\frac{1}{8}$

9. $(15) - (9) = (6); 5\frac{2}{5}$

10. $(2) - (1) = (1); 1\frac{2}{11}$

11. $(8) - (3) = (5); 4\frac{1}{3}$

12. $(16) - (9) = (7); 6\frac{5}{8}$

13. $4\frac{7}{10} - 1\frac{4}{10} = 3\frac{3}{10}$ dollars

14. $1\frac{1}{8} + 1\frac{2}{8} = 2\frac{3}{8}$ mi

Systematic Review 17D

1. $3\frac{3}{7} + 2\frac{1}{7} = 5\frac{4}{7}$

2. $11\frac{3}{4} - 8\frac{2}{4} = 3\frac{1}{4}$

3. $7\frac{1}{5} + 2\frac{3}{5} = 9\frac{4}{5}$

4. $3\frac{4}{5} = \frac{19}{5}$

   (As you may have discovered, a shortcut is to think, "$5 \times 3 = 15$ and $15 + 4 = 19$, so $\frac{19}{5}$.")

5. $4\frac{1}{8} = \frac{33}{8}$

6. $5\frac{5}{6} = \frac{35}{6}$

7. $\frac{19}{9} = 2\frac{1}{9}$

   Remember that a fraction is also a division problem. Dividing and writing the answer with a fractional remainder produces a mixed number.

8. $\frac{24}{5} = 4\frac{4}{5}$

9. $\frac{15}{8} = 1\frac{7}{8}$

10. done

11. $20 \times 20 \times 20 = 8{,}000$ cu ft

12. $12 \times 12 \times 12 = 1{,}728$ cu ft

13. $10 \times 10 \times 10 = 1{,}000$ cu ft

14. $7\frac{2}{3} - 3\frac{1}{3} = 4\frac{1}{3}$ mi

15. $2 \times 3 \times 5 \times 5$

16. $4^2 = 16$

17. $9 \times 10 = 90$ sq ft
    $90 < 100$; yes

18. $\frac{1}{6} \times 18 = 3$ robins

Systematic Review 17E

1. $2\frac{4}{5} - 1\frac{2}{5} = 1\frac{2}{5}$

2. $4\frac{1}{9} + 6\frac{7}{9} = 10\frac{8}{9}$

3. $5\frac{5}{6} - 2\frac{4}{6} = 3\frac{1}{6}$

4. $2\frac{1}{2} = \frac{5}{2}$

5. $6\frac{2}{4} = \frac{26}{4}$

6. $5\frac{3}{8} = \frac{43}{8}$

7. $\frac{51}{10} = 5\frac{1}{10}$

8. $\frac{55}{9} = 6\frac{1}{9}$

9. $\frac{16}{5} = 3\frac{1}{5}$

10. $1\frac{1}{4}$

11. $4 \times 4 \times 4 = 64$ cu in

12. $15 \times 15 \times 15 = 3{,}375$ cu ft

13. $30 \times 30 \times 30 = 27{,}000$ cu ft

14. $1 \times 1 \times 1 = 1$ cu ft

15. $25: \underline{5}, 25$
    $30: 2, 3, \underline{5}, 6, 10, 15, 30$
    GCF = 5

16. $4\frac{1}{4} + 3\frac{2}{4} = 7\frac{3}{4}$ tons

17. $\frac{1}{3} + \frac{2}{3} = \frac{3}{3} = 1$
    $1 + \frac{3}{5} = 1\frac{3}{5}$ yd

18. $\frac{4}{5} \times \frac{1}{2} = \frac{4}{10} = \frac{2}{5}$ of the job

Systematic Review 17F

1. $4\frac{2}{5} + 8\frac{2}{5} = 12\frac{4}{5}$

2. $21\frac{3}{7} - 9\frac{1}{7} = 12\frac{2}{7}$

3. $6\frac{1}{8} + 7\frac{4}{8} = 13\frac{5}{8}$

4. $1\frac{3}{4} = \frac{7}{4}$

5. $2\frac{5}{6} = \frac{17}{6}$

6. $6\frac{1}{3} = \frac{19}{3}$

7. $\frac{5}{4} = 1\frac{1}{4}$

8. $\frac{23}{6} = 3\frac{5}{6}$

9. $\frac{11}{4} = 2\frac{3}{4}$

10. $\frac{1}{2} \times \frac{3}{5} = \frac{3}{10}$

11. $\frac{12}{18} \div \frac{3}{18} = \frac{12 \div 3}{1} = 4$

12. $\frac{4}{7} \times \frac{1}{4} = \frac{4}{28} = \frac{1}{7}$

13. $7 \times 7 \times 7 = 343$ cu in

14. $25 \times 25 \times 25 = 15{,}625$ cu ft

15. $41 \times 41 \times 41 = 68{,}921$ cu ft

16. $2 \times 2 \times 2 = 8$ cu in
    $8 \times 6 = 48$ cu in

17. yes

18. $2\frac{1}{5} + 3\frac{2}{5} = 5\frac{3}{5}$ lb

19. $\frac{1}{2} \times \frac{1}{4} = \frac{1}{8}$ sq mi

20. $\frac{33}{77} > \frac{28}{77}$ so $\frac{3}{7} > \frac{4}{11}$

Lesson Practice 18A

1. done

2. done

3. $1\frac{2}{5} + 4\frac{4}{5} = 5\frac{6}{5} = 6\frac{1}{5}$

4. $3\frac{3}{8} + 1\frac{7}{8} = 4\frac{10}{8} = 5\frac{2}{8} = 5\frac{1}{4}$

5. $7\frac{2}{5} + 2\frac{4}{5} = 9\frac{6}{5} = 10\frac{1}{5}$

6. $2\frac{5}{6} + 4\frac{3}{6} = 6\frac{8}{6} = 7\frac{2}{6} = 7\frac{1}{3}$

7. $4\frac{2}{4} + 2\frac{3}{4} = 6\frac{5}{4} = 7\frac{1}{4}$

8. $5\frac{5}{8} + 7\frac{3}{8} = 12\frac{8}{8} = 13$

9. $2\frac{3}{7} + 5\frac{6}{7} = 7\frac{9}{7} = 8\frac{2}{7}$

10. $3\frac{1}{2} + 3\frac{1}{2} = 6\frac{2}{2} = 7$

11. $7\frac{2}{3} + 3\frac{2}{3} = 10\frac{4}{3} = 11\frac{1}{3}$

12. $4\frac{8}{9} + 8\frac{5}{9} = 12\frac{13}{9} = 13\frac{4}{9}$

Lesson Practice 18B

1. $9\frac{4}{6} + 7\frac{5}{6} = 16\frac{9}{6} = 17\frac{3}{6} = 17\frac{1}{2}$

2. $3\frac{4}{5} + 6\frac{3}{5} = 9\frac{7}{5} = 10\frac{2}{5}$

3. $2\frac{7}{8} + 5\frac{7}{8} = 7\frac{14}{8} = 8\frac{6}{8} = 8\frac{3}{4}$

4. $4\frac{3}{4} + 1\frac{1}{4} = 5\frac{4}{4} = 6$

5. $2\frac{4}{9} + 3\frac{6}{9} = 5\frac{10}{9} = 6\frac{1}{9}$

6. $6\frac{2}{3} + 8\frac{2}{3} = 14\frac{4}{3} = 15\frac{1}{3}$

7. $1\frac{5}{11} + 2\frac{6}{11} = 3\frac{11}{11} = 4$

8. $7\frac{7}{10} + 9\frac{5}{10} = 16\frac{12}{10} = 17\frac{2}{10} = 17\frac{1}{5}$

9. $2\frac{4}{7} + 2\frac{6}{7} = 4\frac{10}{7} = 5\frac{3}{7}$

10. $5\frac{1}{3} + 2\frac{2}{3} = 7\frac{3}{3} = 8$

11. $6\frac{4}{5} + 4\frac{4}{5} = 10\frac{8}{5} = 11\frac{3}{5}$

12. $10\frac{5}{8} + 3\frac{7}{8} = 13\frac{12}{8} = 14\frac{4}{8} = 14\frac{1}{2}$

Lesson Practice 18C

1. $4\frac{5}{8} + 3\frac{3}{8} = 7\frac{8}{8} = 8$

2. $4\frac{5}{6} + 7\frac{4}{6} = 11\frac{9}{6} = 12\frac{3}{6} = 12\frac{1}{2}$

3. $3\frac{8}{9} + 6\frac{7}{9} = 9\frac{15}{9} = 10\frac{6}{9} = 10\frac{2}{3}$

4. $5\frac{4}{5} + 2\frac{2}{5} = 7\frac{6}{5} = 8\frac{1}{5}$

5. $3\frac{5}{10} + 4\frac{7}{10} = 7\frac{12}{10} = 8\frac{2}{10} = 8\frac{1}{5}$

6. $7\frac{3}{4} + 9\frac{2}{4} = 16\frac{5}{4} = 17\frac{1}{4}$

7. $2\frac{6}{12} + 3\frac{7}{12} = 5\frac{13}{12} = 6\frac{1}{12}$

8. $8\frac{8}{11} + 10\frac{6}{11} = 18\frac{14}{11} = 19\frac{3}{11}$

9. $6\frac{5}{8} + 5\frac{7}{8} = 11\frac{12}{8} = 12\frac{4}{8} = 12\frac{1}{2}$

10. $6\frac{2}{4} + 4\frac{2}{4} = 10\frac{4}{4} = 11$

11. $7\frac{5}{6} + 5\frac{4}{6} = 12\frac{9}{6} = 13\frac{3}{6} = 13\frac{1}{2}$

12. $12\frac{5}{9} + 8\frac{8}{9} = 20\frac{13}{9} = 21\frac{4}{9}$

Systematic Review 18D

1. $4\frac{3}{7} + 3\frac{6}{7} = 7\frac{9}{7} = 8\frac{2}{7}$

2. $14\frac{4}{5} + 9\frac{2}{5} = 23\frac{6}{5} = 24\frac{1}{5}$

3. $8\frac{7}{9} + 5\frac{4}{9} = 13\frac{11}{9} = 14\frac{2}{9}$

4. $5\frac{1}{3} = \frac{16}{3}$

5. $7\frac{2}{9} = \frac{65}{9}$

6. $3\frac{4}{5} = \frac{19}{5}$

7. $\frac{27}{5} = 5\frac{2}{5}$

8. $\frac{39}{6} = 6\frac{3}{6} = 6\frac{1}{2}$

9. $\frac{33}{10} = 3\frac{3}{10}$

10. done

11. $12 \times 4 \times 3 = 144$ cu in

12. $2 \times 2 \times 3 = 12$ cu ft

13. $12 \times 13 \times 8 = 1,248$ cu ft

14. $4\frac{5}{8} + 3\frac{7}{8} = 7\frac{12}{8} = 8\frac{4}{8} = 8\frac{1}{2}$ lb

15. 2×37

16. $16 \times 16 = 256$

17. $\left(\frac{1}{2}\right)(10)(15) = 75$ sq ft

18. $3,885 \div 7 = 555$ mi

Systematic Review 18E

1. $2\frac{4}{5} + 1\frac{2}{5} = 3\frac{6}{5} = 4\frac{1}{5}$

2. $4\frac{1}{9} + 6\frac{7}{9} = 10\frac{8}{9}$

3. $5\frac{5}{6} + 2\frac{4}{6} = 7\frac{9}{6} = 8\frac{3}{6} = 8\frac{1}{2}$

4. $\frac{15}{18} - \frac{6}{18} = \frac{9}{18} = \frac{1}{2}$

5. $\frac{16}{24} - \frac{3}{24} = \frac{13}{24}$

6. $\frac{15}{20} + \frac{16}{20} = \frac{31}{20} = 1\frac{11}{20}$

7. $\frac{5}{7} \times \frac{1}{2} = \frac{5}{14}$

8. $\frac{12}{20} \div \frac{15}{20} = \frac{12 \div 15}{1} = \frac{12}{15} = \frac{4}{5}$

9. $\frac{1}{4} \times \frac{5}{6} = \frac{5}{24}$

10. $3\frac{10}{16} = 3\frac{5}{8}$

11. $12 \times 14 \times 5 = 840$ cu in

12. $21 \times 9 \times 8 = 1,512$ cu in

13. $7 \times 6 \times 10 = 420$ cu ft

14. $20 \times 15 \times 6 = 1,800$ cu ft

15. $1,800 \times 63 = 113,400$ lb

16. $16 : \underline{2}, \underline{4}, 8, 16$
    $20 : \underline{2}, \underline{4}, 5, 10, 20$
    GCF = 4

17. $32 + 32 = 64$ ft

18. $5\frac{2}{3} + \frac{1}{3} = 5\frac{3}{3} = 6$ ft

Systematic Review 18F

1. $5\frac{4}{6} + 9\frac{2}{6} = 14\frac{6}{6} = 15$

2. $11\frac{5}{8} + 3\frac{7}{8} = 14\frac{12}{8} = 15\frac{4}{8} = 15\frac{1}{2}$

3. $7\frac{7}{10} + 8\frac{9}{10} = 15\frac{16}{10} = 16\frac{6}{10} = 16\frac{3}{5}$

4. $\frac{45}{63} - \frac{14}{63} = \frac{31}{63}$

5. $\frac{5}{40} + \frac{32}{40} = \frac{37}{40}$

6. $\frac{48}{56} - \frac{7}{56} = \frac{41}{56}$

7. $\frac{40}{48} \div \frac{42}{48} = \frac{40 \div 42}{1} = \frac{40}{42} = \frac{20}{21}$

8. $\frac{1}{2} \times \frac{3}{5} = \frac{3}{10}$

9. $\frac{12}{18} \div \frac{3}{18} = \frac{12 \div 3}{1} = 4$

10. $56 \div 7 = 8$; $8 \times 4 = 32$

11. $50 \div 5 = 10$; $10 \times 3 = 30$

12. $14 \div 2 = 7$; $7 \times 1 = 7$

13. $10 \times 15 \times 7 = 1,050$ cu in

14. $33 \times 19 \times 12 = 7,524$ cu in

15. $5 \times 4 \times 8 = 160$ cu ft

16. $25 \times 13 \times 8 = 2,600$ cu ft full
    $2,600 \div 2 = 1,300$ cu ft half full

17. $1,300 \times 63 = 81,900$ lb

18. yes

19. $\dfrac{1}{2} + \dfrac{1}{2} = \dfrac{2}{2} = 1$

$\dfrac{1}{4} + \dfrac{1}{4} = \dfrac{2}{4} = \dfrac{1}{2}$

$1 + \dfrac{1}{2} = 1\dfrac{1}{2}$ mi

20. $\dfrac{4}{5} \div \dfrac{4}{1} = \dfrac{4}{5} \div \dfrac{20}{5} =$

$\dfrac{4 \div 20}{1} = \dfrac{4}{20} = \dfrac{1}{5}$ of a pie

Lesson Practice 19A

1. done

2. done

3. $6\dfrac{1}{5} \qquad 5\dfrac{6}{5}$

   $-2\dfrac{3}{5} \qquad -2\dfrac{3}{5}$

   $\qquad\qquad 3\dfrac{3}{5}$

4. $4\dfrac{1}{8} \qquad 3\dfrac{9}{8}$

   $-1\dfrac{5}{8} \qquad -1\dfrac{5}{8}$

   $\qquad\qquad 2\dfrac{4}{8} = 2\dfrac{1}{2}$

5. $2\dfrac{1}{4} \qquad 1\dfrac{5}{4}$

   $-1\dfrac{2}{4} \qquad -1\dfrac{2}{4}$

   $\qquad\qquad \dfrac{3}{4}$

6. $6 \qquad 5\dfrac{8}{8}$

   $-3\dfrac{7}{8} \qquad -3\dfrac{7}{8}$

   $\qquad\qquad 2\dfrac{1}{8}$

7. $7\dfrac{5}{16} \qquad 6\dfrac{21}{16}$

   $-2\dfrac{7}{16} \qquad -2\dfrac{7}{16}$

   $\qquad\qquad 4\dfrac{14}{16} = 4\dfrac{7}{8}$

8. $8\dfrac{3}{5}$

   $-2\dfrac{3}{5}$

   $\qquad 6$

9. $4\dfrac{1}{6} \qquad 3\dfrac{7}{6}$

   $-1\dfrac{5}{6} \qquad -1\dfrac{5}{6}$

   $\qquad\qquad 2\dfrac{2}{6} = 2\dfrac{1}{3}$

10. $9\dfrac{1}{8} \qquad 8\dfrac{9}{8}$

    $-2\dfrac{7}{8} \qquad -2\dfrac{7}{8}$

    $\qquad\qquad 6\dfrac{2}{8} = 6\dfrac{1}{4}$

11. $4 \qquad 3\dfrac{3}{3}$

    $-\dfrac{1}{3} \qquad -\dfrac{1}{3}$

    $\qquad\qquad 3\dfrac{2}{3}$

12. $7\dfrac{2}{5} \qquad 6\dfrac{7}{5}$

    $-6\dfrac{4}{5} \qquad -6\dfrac{4}{5}$

    $\qquad\qquad \dfrac{3}{5}$

Lesson Practice 19B

1. $14\dfrac{3}{7} \qquad 13\dfrac{10}{7}$

   $-2\dfrac{4}{7} \qquad -2\dfrac{4}{7}$

   $\qquad\qquad 11\dfrac{6}{7}$

2. $11 \qquad 10\dfrac{9}{9}$

   $-2\dfrac{8}{9} \qquad -2\dfrac{8}{9}$

   $\qquad\qquad 8\dfrac{1}{9}$

3. $8\frac{3}{10}$ $7\frac{13}{10}$

     $-5\frac{4}{10}$ $-5\frac{4}{10}$

            $2\frac{9}{10}$

4. $19\frac{2}{6}$ $18\frac{8}{6}$

     $-10\frac{3}{6}$ $-10\frac{3}{6}$

            $8\frac{5}{6}$

5. $3\frac{1}{3}$ $2\frac{4}{3}$

     $-1\frac{2}{3}$ $-1\frac{2}{3}$

            $1\frac{2}{3}$

6. 5 $4\frac{7}{7}$

     $-1\frac{4}{7}$ $-1\frac{4}{7}$

            $3\frac{3}{7}$

7. $5\frac{1}{4}$ $4\frac{5}{4}$

     $-2\frac{3}{4}$ $-2\frac{3}{4}$

            $2\frac{2}{4}$ $= 2\frac{1}{2}$

8. $6\frac{1}{5}$ $5\frac{6}{5}$

     $-2\frac{3}{5}$ $-2\frac{3}{5}$

            $3\frac{3}{5}$

9. $4\frac{1}{8}$ $3\frac{9}{8}$

     $-1\frac{5}{8}$ $-1\frac{5}{8}$

            $2\frac{4}{8}$ $= 2\frac{1}{2}$

10. $6\frac{5}{8}$ $5\frac{13}{8}$

     $-3\frac{7}{8}$ $-3\frac{7}{8}$

            $2\frac{6}{8}$ $= 2\frac{3}{4}$

11. 8 $7\frac{10}{10}$

     $-\frac{3}{10}$ $-\frac{3}{10}$

            $7\frac{7}{10}$

12. $2\frac{1}{4}$ $1\frac{5}{4}$

     $-1\frac{3}{4}$ $-1\frac{3}{4}$

            $\frac{2}{4}$ $= \frac{1}{2}$

Lesson Practice 19C

1. $5\frac{1}{3}$ $\rightarrow$ $4\frac{4}{3}$

     $-2\frac{2}{3}$ $-2\frac{2}{3}$

            $2\frac{2}{3}$

2. 6 $\rightarrow$ $5\frac{8}{8}$

     $-2\frac{3}{8}$ $-2\frac{3}{8}$

            $3\frac{5}{8}$

3. $6\frac{3}{16}$ $\rightarrow$ $5\frac{19}{16}$

     $-2\frac{5}{16}$ $-2\frac{5}{16}$

            $3\frac{14}{16}$ $= 3\frac{7}{8}$

4. $8\frac{1}{5}$ $\rightarrow$ $7\frac{6}{5}$

     $-2\frac{4}{5}$ $-2\frac{4}{5}$

            $5\frac{2}{5}$

5. $9\frac{1}{4}$ $\rightarrow$ $8\frac{5}{4}$

     $-6\frac{3}{4}$ $-6\frac{3}{4}$

            $2\frac{2}{4}$ $= 2\frac{1}{2}$

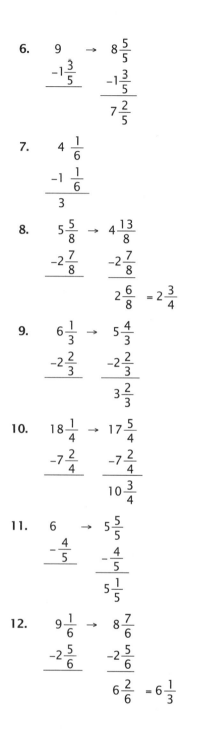

6. $9 \rightarrow 8\frac{5}{5}$
   $-1\frac{3}{5} \quad -1\frac{3}{5}$
   $\overline{\qquad 7\frac{2}{5}}$

7. $4\frac{1}{6}$
   $-1\frac{1}{6}$
   $\overline{\quad 3}$

8. $5\frac{5}{8} \rightarrow 4\frac{13}{8}$
   $-2\frac{7}{8} \quad -2\frac{7}{8}$
   $\overline{\qquad 2\frac{6}{8} = 2\frac{3}{4}}$

9. $6\frac{1}{3} \rightarrow 5\frac{4}{3}$
   $-2\frac{2}{3} \quad -2\frac{2}{3}$
   $\overline{\qquad 3\frac{2}{3}}$

10. $18\frac{1}{4} \rightarrow 17\frac{5}{4}$
    $-7\frac{2}{4} \quad -7\frac{2}{4}$
    $\overline{\qquad 10\frac{3}{4}}$

11. $6 \rightarrow 5\frac{5}{5}$
    $-\frac{4}{5} \quad -\frac{4}{5}$
    $\overline{\qquad 5\frac{1}{5}}$

12. $9\frac{1}{6} \rightarrow 8\frac{7}{6}$
    $-2\frac{5}{6} \quad -2\frac{5}{6}$
    $\overline{\qquad 6\frac{2}{6} = 6\frac{1}{3}}$

Systematic Review 19D

1. $5\frac{1}{5} \rightarrow 4\frac{6}{5}$
   $-2\frac{3}{5} \quad -2\frac{3}{5}$
   $\overline{\qquad 2\frac{3}{5}}$

2. $8\frac{5}{8} \rightarrow 7\frac{13}{8}$
   $-2\frac{7}{8} \quad -2\frac{7}{8}$
   $\overline{\qquad 5\frac{6}{8} = 5\frac{3}{4}}$

3. $6 \rightarrow 5\frac{3}{3}$
   $-4\frac{2}{3} \quad -4\frac{2}{3}$
   $\overline{\qquad 1\frac{1}{3}}$

4. $7\frac{1}{5}$
   $+ 2\frac{4}{5}$
   $\overline{9\frac{5}{5} = 10}$

5. $1\frac{3}{4}$
   $+ 2\frac{3}{4}$
   $\overline{3\frac{6}{4} = 4\frac{2}{4} = 4\frac{1}{2}}$

6. $9\frac{7}{12}$
   $+ 3\frac{11}{12}$
   $\overline{12\frac{18}{12} = 13\frac{6}{12} = 13\frac{1}{2}}$

7. $\frac{1}{6} \times \frac{4}{5} = \frac{4}{30} = \frac{2}{15}$

8. $\frac{3}{6} \times \frac{1}{2} = \frac{3}{12} = \frac{1}{4}$

9. $\frac{4}{5} \times \frac{2}{7} = \frac{8}{35}$

10. done

11. done

12. $6 \div 3 = 2$ yd

13. $5 \times 3 = 15$ ft

14. $3 \times 3 = 9$ ft

$9 \times 15 = 135$ sq ft

15. $15 \div 3 = 5$ yd

$5 \times 3 = 15$ sq yd

16. $\dfrac{3}{8} + \dfrac{1}{8} = \dfrac{4}{8}$

$\dfrac{4}{8} + \dfrac{3}{4} = \dfrac{16}{32} + \dfrac{24}{32} =$

$\dfrac{40}{32} = 1\dfrac{8}{32} = 1\dfrac{1}{4}$ ft

17. $4\dfrac{5}{4} - 3\dfrac{3}{4} = 1\dfrac{2}{4} = 1\dfrac{1}{2}$ ft

18. $3 \div 3 = 1$ yd

$1 \times 1 \times 1 = 1$ cu yd

Systematic Review 19E

1.
$$7\dfrac{3}{6} \quad \rightarrow \quad 6\dfrac{9}{6}$$
$$-2\dfrac{5}{6} \qquad\quad -2\dfrac{5}{6}$$
$$\qquad\qquad\qquad 4\dfrac{4}{6} = 4\dfrac{2}{3}$$

2.
$$6\dfrac{4}{7} \quad \rightarrow \quad 5\dfrac{11}{7}$$
$$-\dfrac{6}{7} \qquad\quad -\dfrac{6}{7}$$
$$\qquad\qquad\qquad 5\dfrac{5}{7}$$

3.
$$5 \quad \rightarrow \quad 4\dfrac{8}{8}$$
$$-4\dfrac{6}{8} \qquad\quad -4\dfrac{6}{8}$$
$$\qquad\qquad\qquad \dfrac{2}{8} = \dfrac{1}{4}$$

4.
$$2\dfrac{1}{2}$$
$$+ 3\dfrac{1}{2}$$
$$5\dfrac{2}{2} = 6$$

5.
$$8\dfrac{2}{7}$$
$$+ 8\dfrac{6}{7}$$
$$16\dfrac{8}{7} = 17\dfrac{1}{7}$$

6.
$$5\dfrac{4}{7}$$
$$+ 2\dfrac{5}{7}$$
$$7\dfrac{9}{7} = 8\dfrac{2}{7}$$

7. $\dfrac{7}{8} \div \dfrac{1}{8} = \dfrac{7 \div 1}{1} = 7$

8. $\dfrac{15}{27} \div \dfrac{18}{27} = \dfrac{15 \div 18}{1} = \dfrac{15}{18} = \dfrac{5}{6}$

9. $\dfrac{4}{12} \div \dfrac{9}{12} = \dfrac{4 \div 9}{1} = \dfrac{4}{9}$

10. $1\dfrac{2}{10} = 1\dfrac{1}{5}$

11. $24 \div 3 = 8$ yd

12. $10 \times 3 = 30$ ft

13. $36 \div 3 = 12$ yd

14. $16 \times 3 = 48$ ft

15. $\dfrac{1}{2} \times \dfrac{2}{5} = \dfrac{2}{10} = \dfrac{1}{5}$

$\dfrac{1}{5} \times \dfrac{1}{2} = \dfrac{1}{10}$ sq mi

16. $\left(\dfrac{1}{2}\right)(20)(4) = 40$ cu ft

17. $3\dfrac{1}{8} + 4\dfrac{5}{8} = 7\dfrac{6}{8}$

$7\dfrac{6}{8} - 2\dfrac{7}{8} =$

$6\dfrac{14}{8} - 2\dfrac{7}{8} = 4\dfrac{7}{8}$ lb

18. $125 \div 25 = 5$ ft

Systematic Review 19F

1.
$$25\frac{1}{3}$$
$$-12\frac{1}{3}$$
$$\overline{\quad 13 \quad}$$

2.
$$1\frac{5}{8} \;\rightarrow\; \frac{13}{8}$$
$$-\frac{6}{8} \qquad -\frac{6}{8}$$
$$\overline{\qquad\qquad \frac{7}{8}}$$

3.
$$9 \;\rightarrow\; 8\frac{7}{7}$$
$$-4\frac{2}{7} \qquad -4\frac{2}{7}$$
$$\overline{\qquad\qquad 4\frac{5}{7}}$$

4.
$$1\frac{3}{4}$$
$$+\,2\frac{1}{4}$$
$$\overline{3\frac{4}{4}} = 4$$

5.
$$10\frac{2}{6}$$
$$+\,7\frac{5}{6}$$
$$\overline{17\frac{7}{6}} = 18\frac{1}{6}$$

6.
$$9\frac{5}{9}$$
$$+\,5\frac{8}{9}$$
$$\overline{14\frac{13}{9}} = 15\frac{4}{9}$$

7. $\dfrac{7}{14}+\dfrac{6}{14}=\dfrac{13}{14}$

$\dfrac{13}{14}+\dfrac{4}{9}=\dfrac{117}{126}+\dfrac{56}{126}=\dfrac{173}{126}=1\dfrac{47}{126}$

8. $\dfrac{3}{6}+\dfrac{4}{6}=\dfrac{7}{6}$

$\dfrac{7}{6}+\dfrac{3}{5}=\dfrac{35}{30}+\dfrac{18}{30}=\dfrac{53}{30}=1\dfrac{23}{30}$

9. $\dfrac{8}{12}+\dfrac{3}{12}+\dfrac{1}{12}=\dfrac{12}{12}=1$

10. $\dfrac{15}{20}<\dfrac{16}{20}$

11. $\dfrac{32}{72}>\dfrac{27}{72}$

12. $\dfrac{72}{132}>\dfrac{55}{132}$

13. $45 \div 3 = 15$ yd

14. $17 \times 3 = 51$ ft

15. $3 \times 3 = 9$ ft

16. $\dfrac{5}{8}\times\dfrac{1}{5}=\dfrac{5}{40}=\dfrac{1}{8}$ sq mi

17. $\dfrac{5}{8}+\dfrac{5}{8}=\dfrac{10}{8}=\dfrac{5}{4};\ \dfrac{1}{5}+\dfrac{1}{5}=\dfrac{2}{5}$

$\dfrac{5}{4}+\dfrac{2}{5}=\dfrac{25}{20}+\dfrac{8}{20}=\dfrac{33}{20}=1\dfrac{13}{20}$ mi

18. $2\dfrac{3}{5}-1\dfrac{4}{5}=1\dfrac{8}{5}-1\dfrac{4}{5}=\dfrac{4}{5}$ bushel

19. $30 \times 3 = 90$ ft

20. $1{,}260 \div 45 = 28$ ft

Lesson Practice 20A

1. done

2. done

3.
$$4\frac{1}{3}+\frac{1}{3}=4\frac{2}{3}$$
$$-3\frac{2}{3}+\frac{1}{3}=4$$
$$\overline{\qquad\qquad \frac{2}{3}}$$

4.
$$7+\frac{1}{4}=7\frac{1}{4}$$
$$-2\frac{3}{4}+\frac{1}{4}=3$$
$$\overline{\qquad\qquad 4\frac{1}{4}}$$

5.
$$6\frac{1}{5}+\frac{3}{5}=6\frac{4}{5}$$
$$-3\frac{2}{5}+\frac{3}{5}=4$$
$$\overline{\qquad\qquad 2\frac{4}{5}}$$

6.
$$8+\frac{3}{8}=8\frac{3}{8}$$
$$-1\frac{5}{8}+\frac{3}{8}=2$$
$$\overline{\qquad\qquad 6\frac{3}{8}}$$

7. $9\frac{5}{16} + \frac{9}{16} = 9\frac{14}{16}$

$\underline{-2\frac{7}{16} + \frac{9}{16} = 3}$

$= 6\frac{14}{16} = 6\frac{7}{8}$

8. $10 + \frac{1}{8} = 10\frac{1}{8}$

$\underline{-4\frac{7}{8} + \frac{1}{8} = 5}$

$5\frac{1}{8}$

Lesson Practice 20B

1. $5\frac{2}{5} + \frac{1}{5} = 5\frac{3}{5}$

$\underline{-1\frac{4}{5} + \frac{1}{5} = 2}$

$3\frac{3}{5}$

2. $9 + \frac{1}{4} = 9\frac{1}{4}$

$\underline{-2\frac{3}{4} + \frac{1}{4} = 3}$

$6\frac{1}{4}$

3. $10\frac{1}{8} + \frac{1}{8} = 10\frac{2}{8}$

$\underline{-2\frac{7}{8} + \frac{1}{8} = 3}$

$= 7\frac{2}{8} = 7\frac{1}{4}$

4. $7 + \frac{2}{3} = 7\frac{2}{3}$

$\underline{-6\frac{1}{3} + \frac{2}{3} = 7}$

$\frac{2}{3}$

5. $12\frac{1}{6} + \frac{4}{6} = 12\frac{5}{6}$

$\underline{-8\frac{2}{6} + \frac{4}{6} = 9}$

$3\frac{5}{6}$

6. $25 + \frac{2}{2} = 25\frac{2}{5}$

$\underline{-5\frac{3}{5} + \frac{2}{5} = 6}$

$19\frac{2}{5}$

7. $8\frac{3}{10} + \frac{1}{10} = 8\frac{4}{10}$

$\underline{-1\frac{9}{10} + \frac{1}{10} = 2}$

$= 6\frac{4}{10} = 6\frac{2}{5}$

8. $4 + \frac{3}{8} = 4\frac{3}{8}$

$\underline{-3\frac{5}{8} + \frac{3}{8} = 4}$

$\frac{3}{8}$

Lesson Practice 20C

1. $6\frac{3}{6} + \frac{1}{6} = 6\frac{4}{6}$

$\underline{-2\frac{5}{6} + \frac{1}{6} = 3}$

$= 3\frac{4}{6} = 3\frac{2}{3}$

2. $14 + \frac{1}{5} = 14\frac{1}{5}$

$\underline{-5\frac{4}{5} + \frac{1}{5} = 6}$

$8\frac{1}{5}$

3. $16\frac{2}{9} + \frac{5}{9} = 16\frac{7}{9}$

$\underline{-7\frac{4}{9} + \frac{5}{9} = 8}$

$8\frac{7}{9}$

4. $9 + \frac{1}{4} = 9\frac{1}{4}$

$\underline{-3\frac{3}{4} + \frac{1}{4} = 4}$

$5\frac{1}{4}$

5. $13\frac{3}{7} + \frac{1}{7} = 13\frac{4}{7}$

    $\underline{-\ 8\frac{6}{7} + \frac{1}{7} = \ 9}$

    $\qquad\qquad\qquad 4\frac{4}{7}$

6. $30 + \frac{3}{10} = 30\frac{3}{10}$

    $\underline{-11\frac{7}{10} + \frac{3}{10} = 12}$

    $\qquad\qquad\qquad 18\frac{3}{10}$

7. $6\frac{1}{4} + \frac{2}{4} = 6\frac{3}{4}$

    $\underline{-\ 2\frac{2}{4} + \frac{2}{4} = 3}$

    $\qquad\qquad\qquad 3\frac{3}{4}$

8. $5 + \frac{5}{8} = 5\frac{5}{8}$

    $\underline{-\ 1\frac{3}{8} + \frac{5}{8} = 2}$

    $\qquad\qquad\qquad 3\frac{5}{8}$

Systematic Review 20D

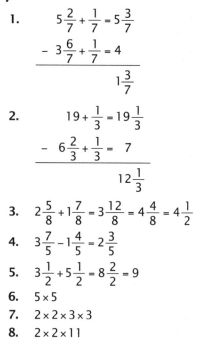

1. $5\frac{2}{7} + \frac{1}{7} = 5\frac{3}{7}$

    $\underline{-\ 3\frac{6}{7} + \frac{1}{7} = 4}$

    $\qquad\qquad\qquad 1\frac{3}{7}$

2. $19 + \frac{1}{3} = 19\frac{1}{3}$

    $\underline{-\ 6\frac{2}{3} + \frac{1}{3} = \ 7}$

    $\qquad\qquad\qquad 12\frac{1}{3}$

3. $2\frac{5}{8} + 1\frac{7}{8} = 3\frac{12}{8} = 4\frac{4}{8} = 4\frac{1}{2}$

4. $3\frac{7}{5} - 1\frac{4}{5} = 2\frac{3}{5}$

5. $3\frac{1}{2} + 5\frac{1}{2} = 8\frac{2}{2} = 9$

6. 5×5

7. $2 \times 2 \times 3 \times 3$

8. $2 \times 2 \times 11$

9. $\frac{3}{7} = \frac{6}{14} = \frac{9}{21}$

10. $\frac{2}{3} = \frac{4}{6} = \frac{6}{9}$

11. $\frac{1}{9} = \frac{2}{18} = \frac{3}{27}$

12. done

13. done

14. $12 \div 2 = 6$ qt

15. $11 \times 2 = 22$ pt

16. $19\frac{2}{2} - 5\frac{1}{2} = 14\frac{1}{2}$ lb

17. $7 \times 2 = 14$ pint jars

18. $6 \times 3 = 18$ ft

    $18 \times 18 \times 18 = 5,832$ cu ft

Systematic Review 20E

1. $7\frac{4}{9} + \frac{2}{9} = 7\frac{6}{9}$

    $\underline{-\ 3\frac{7}{9} + \frac{2}{9} = 4}$

    $\qquad\qquad = 3\frac{6}{9} = 3\frac{2}{3}$

2. $13 + \frac{7}{8} = 13\frac{7}{8}$

    $\underline{-\ 7\frac{1}{8} + \frac{7}{8} = \ 8}$

    $\qquad\qquad\qquad 5\frac{7}{8}$

3. $2\frac{1}{5} + 4\frac{3}{5} = 6\frac{4}{5}$

4. $5\frac{9}{6} - 4\frac{5}{6} = 1\frac{4}{6} = 1\frac{2}{3}$

5. $5\frac{2}{3} + 5\frac{2}{3} = 10\frac{4}{3} = 11\frac{1}{3}$

6. $25 : \underline{5}, 25$

    $35 : \underline{5}, 7, 35$

    GCF $= 5$

7. $12 : \underline{2}, \underline{3}, \underline{4}, \underline{6}, \underline{12}$

    $36 : \underline{2}, \underline{3}, \underline{4}, \underline{6}, 9, \underline{12}, 18, 36$

    GCF $= 12$

8. $42 : 2, 3, 6, \underline{7}, 14, 21, 42$

    $49 : \underline{7}, 49$

    GCF $= 7$

9. $\frac{6}{12} - \frac{2}{12} = \frac{4}{12} = \frac{1}{3}$

10. $\dfrac{35}{50} - \dfrac{20}{50} = \dfrac{15}{50} = \dfrac{3}{10}$

11. $\dfrac{99}{108} - \dfrac{60}{108} = \dfrac{39}{108} = \dfrac{13}{36}$

12. $11 \times 2 = 22$ pt

13. $18 \div 2 = 9$ qt

14. $22 \div 2 = 11$ qt

15. $2\dfrac{5}{8} + 3\dfrac{7}{8} = 5\dfrac{12}{8} = 6\dfrac{4}{8} = 6\dfrac{1}{2}$ in

16. no

17. $16 \div 2 = 8$ qt

18. Answers will vary.

19. $25 + 43 = 68$

$\dfrac{1}{4} \times 68 = \$17$

20. $7 \times 7 = 49$

Systematic Review 20F

1. $9\dfrac{1}{6} + \dfrac{1}{6} = 9\dfrac{2}{6}$

$\underline{-\ 4\dfrac{5}{6} + \dfrac{1}{6} = 5}$

$= 4\dfrac{2}{6} = 4\dfrac{1}{3}$

2. $18 + \dfrac{1}{2} = 18\dfrac{1}{2}$

$\underline{-\ 12\dfrac{1}{2} + \dfrac{1}{2} = 13}$

$5\dfrac{1}{2}$

3. $3\dfrac{3}{8} + 5\dfrac{5}{8} = 8\dfrac{8}{8} = 9$

4. $8\dfrac{1}{3} - 6\dfrac{1}{3} = 2$

5. $7\dfrac{5}{9} + 7\dfrac{8}{9} = 14\dfrac{13}{9} = 15\dfrac{4}{9}$

6. $192 \div 3 = 64;\ 64 \times 2 = 128$

7. $555 \div 5 = 111;\ 111 \times 1 = 111$

8. $84 \div 4 = 21;\ 21 \times 3 = 63$

9. $\dfrac{6}{8} + \dfrac{4}{8} = \dfrac{10}{8} = \dfrac{5}{4} = 1\dfrac{1}{4}$

10. $\dfrac{40}{110} + \dfrac{33}{110} = \dfrac{73}{110}$

11. $\dfrac{15}{27} + \dfrac{9}{27} = \dfrac{24}{27} = \dfrac{8}{9}$

12. $12 \times 2 = 24$ pt

13. $28 \div 2 = 14$ qt

14. $15 \times 3 = 45$ ft

15. $4\dfrac{2}{10} - 1\dfrac{9}{10} = 3\dfrac{12}{10} - 1\dfrac{9}{10} = 2\dfrac{3}{10}$ in

16. yes

17. $4 + 6 = 10$ qt of juice

$10 \div 5 = 2$ qt per person

$2 \times 2 = 4$ pt per person

18. 11 in

19. $1\dfrac{1}{2} + 1\dfrac{1}{2} + 1\dfrac{1}{2} + 1\dfrac{1}{2} = 4\dfrac{4}{2} = 6$ ft

20. $\dfrac{1}{8} \div \dfrac{1}{4} = \dfrac{4}{32} \div \dfrac{8}{32} = \dfrac{4 \div 8}{1} = \dfrac{4}{8} = \dfrac{1}{2}$ mi

Lesson Practice 21A

1. done

2. $3\dfrac{15}{40} + 1\dfrac{32}{40} = 4\dfrac{47}{40} = 5\dfrac{7}{40}$

3. $18\dfrac{32}{80} + 3\dfrac{50}{80} = 21\dfrac{82}{80} = 22\dfrac{2}{80} = 22\dfrac{1}{40}$

4. $11\dfrac{18}{30} + 4\dfrac{25}{30} = 15\dfrac{43}{30} = 16\dfrac{13}{30}$

5. $9\dfrac{4}{12} + 6\dfrac{3}{12} = 15\dfrac{7}{12}$

6. $4\dfrac{10}{15} + 1\dfrac{6}{15} = 5\dfrac{16}{15} = 6\dfrac{1}{15}$

7. $9\dfrac{30}{50} + 2\dfrac{35}{50} = 11\dfrac{65}{50} = 12\dfrac{15}{50} = 12\dfrac{3}{10}$

8. $12\dfrac{72}{80} + 4\dfrac{50}{80} = 16\dfrac{122}{80} = 16\dfrac{61}{40} = 17\dfrac{21}{40}$

9. $3\dfrac{4}{8} + 1\dfrac{6}{8} = 4\dfrac{10}{8} = 5\dfrac{2}{8} = 5\dfrac{1}{4}$

10. $4\dfrac{20}{32} + 1\dfrac{8}{32} = 5\dfrac{28}{32} = 5\dfrac{7}{8}$

Lesson Practice 21B

1. $7\frac{15}{20} + 9\frac{12}{20} = 16\frac{27}{20} = 17\frac{7}{20}$

2. $6\frac{18}{24} + 6\frac{20}{24} = 12\frac{38}{24} = 12\frac{19}{12} = 13\frac{7}{12}$

3. $19\frac{70}{80} + 2\frac{40}{80} = 21\frac{110}{80} = 21\frac{11}{8} = 22\frac{3}{8}$

4. $12\frac{24}{33} + 5\frac{22}{33} = 17\frac{46}{33} = 18\frac{13}{33}$

5. $8\frac{12}{15} + 2\frac{5}{15} = 10\frac{17}{15} = 11\frac{2}{15}$

6. $4\frac{15}{20} + 2\frac{4}{20} = 6\frac{19}{20}$

7. $6\frac{15}{27} + 8\frac{9}{27} = 14\frac{24}{27} = 14\frac{8}{9}$

8. $6\frac{18}{27} + 1\frac{21}{27} = 7\frac{39}{27} = 7\frac{13}{9} = 8\frac{4}{9}$

9. $7\frac{15}{50} + 9\frac{40}{50} = 16\frac{55}{50} = 16\frac{11}{10} = 17\frac{1}{10}$

10. $2\frac{6}{10} + 1\frac{5}{10} = 3\frac{11}{10} = 4\frac{1}{10}$

Lesson Practice 21C

1. $6\frac{4}{10} + 2\frac{5}{10} = 8\frac{9}{10}$

2. $5\frac{27}{63} + 3\frac{7}{63} = 8\frac{34}{63}$

3. $2\frac{5}{40} + \frac{32}{40} = 2\frac{37}{40}$

4. $8\frac{27}{63} + 7\frac{14}{63} = 15\frac{41}{63}$

5. $3\frac{8}{10} + 4\frac{5}{10} = 7\frac{13}{10} = 8\frac{3}{10}$

6. $2\frac{7}{21} + 5\frac{15}{21} = 7\frac{22}{21} = 8\frac{1}{21}$

7. $6\frac{6}{10} + 4\frac{5}{10} = 10\frac{11}{10} = 11\frac{1}{10}$

8. $1\frac{15}{20} + 5\frac{12}{20} = 6\frac{27}{20} = 7\frac{7}{20}$

9. $13\frac{12}{42} + 4\frac{7}{42} = 17\frac{19}{42}$

10. $16\frac{4}{8} + 17\frac{6}{8} = 33\frac{10}{8} = 34\frac{2}{8} = 34\frac{1}{4}$

Systematic Review 21D

1. $8\frac{25}{35} + 4\frac{21}{35} = 12\frac{46}{35} = 13\frac{11}{35}$

2. $6\frac{20}{28} + 1\frac{21}{28} = 7\frac{41}{28} = 8\frac{13}{28}$

3. $4\frac{7}{6} - 2\frac{4}{6} = 2\frac{3}{6} = 2\frac{1}{2}$

4. $6\frac{2}{5} - 1\frac{1}{5} = 5\frac{1}{5}$

5. $13\frac{13}{10} - 9\frac{9}{10} = 4\frac{4}{10} = 4\frac{2}{5}$

6. $\frac{2}{6} \div \frac{3}{6} = \frac{2 \div 3}{1} = \frac{2}{3}$

7. $\frac{9}{18} \div \frac{12}{18} = \frac{9 \div 12}{1} = \frac{9}{12} = \frac{3}{4}$

8. $\frac{20}{50} \div \frac{20}{50} = \frac{20 \div 20}{1} = 1$

9. $\frac{1}{8} \times \frac{6}{7} = \frac{6}{56} = \frac{3}{28}$

10. $\frac{4}{5} \times \frac{1}{9} = \frac{4}{45}$

11. $\frac{3}{5} \times \frac{5}{6} = \frac{15}{30} = \frac{1}{2}$

12. $99 \div 3 = 33$ yd

13. $6 \div 2 = 3$ qt

14. $36 \times 2 = 72$ pt

15. $5 + 4 = 9$
    $9 \times 8 = 72$
    $72 - 2 = 70$
    $70 \div 10 = 7$
    $7 + 3 = 10$

16. $7 - 3 = 4$
    $4 \times 6 = 24$
    $24 \div 3 = 8$
    $8 \times 9 = 72$

17. $13 \div 3 = 4\frac{1}{3}$ yd

18. $1\frac{4}{8} + 1\frac{6}{8} = 2\frac{10}{8} = 2\frac{5}{4} = 3\frac{1}{4}$ hr

Systematic Review 21E

1. $8\frac{8}{12} + 5\frac{9}{12} = 13\frac{17}{12} = 14\frac{5}{12}$

2. $3\frac{15}{40} + 1\frac{16}{40} = 4\frac{31}{40}$

3. $2\frac{5}{4} - 1\frac{3}{4} = 1\frac{2}{4} = 1\frac{1}{2}$

4. $6\frac{6}{5} - 6\frac{4}{5} = \frac{2}{5}$

5. $4\frac{3}{3} - 1\frac{2}{3} = 3\frac{1}{3}$

6. $\frac{20}{32} \div \frac{8}{32} = \frac{20 \div 8}{1} = \frac{20}{8} = \frac{5}{2} = 2\frac{1}{2}$

7. $\frac{3}{7} \div \frac{1}{7} = \frac{3 \div 1}{1} = 3$

8. $\frac{15}{27} \div \frac{18}{27} = \frac{15 \div 18}{1} = \frac{15}{18} = \frac{5}{6}$

9. $\frac{1}{9} \times \frac{7}{8} = \frac{7}{72}$

10. $\frac{2}{3} \times \frac{1}{2} = \frac{2}{6} = \frac{1}{3}$

11. $\frac{7}{10} \times \frac{1}{5} = \frac{7}{50}$

12. $7\frac{1}{2} + 7\frac{1}{2} = 14\frac{2}{2} = 15$

    $3\frac{1}{4} + 3\frac{1}{4} = 6\frac{2}{4} = 6\frac{1}{2}$

    $15 + 6\frac{1}{2} = 21\frac{1}{2}$ in

13. $5\frac{1}{2} + 5\frac{1}{2} + 5\frac{1}{2} + 5\frac{1}{2} = 20\frac{4}{2} = 22$ in

14. $4\frac{1}{3} + 5\frac{2}{3} = 9\frac{3}{3} = 10$

    $10 + 3\frac{1}{2} = 13\frac{1}{2}$ ft

15. $8 \times 6 \times 3 = 144$ cu ft

16. $144 \div 12 = 12$ cu ft drained

    $144 - 12 = 132$ cu ft left

17. $17,259 \div 3 = 5,753$ yd

18. $25 \div 2 = 12\frac{1}{2}$ qt

19. Answers will vary.

20. $21 \div 7 = 3$

    $3 \times 4 = 12$

    $12 + 3 = 15$

    $15 - 10 = 5$

Systematic Review 21F

1. $18\frac{56}{80} + 3\frac{50}{80} = 21\frac{106}{80} = 22\frac{26}{80} = 22\frac{13}{40}$

2. $11\frac{18}{30} + 4\frac{5}{30} = 15\frac{23}{30}$

3. $9\frac{3}{7} - 5\frac{1}{7} = 4\frac{2}{7}$

4. $16\frac{7}{5} - 8\frac{3}{5} = 8\frac{4}{5}$

5. $5\frac{8}{8} - 2\frac{7}{8} = 3\frac{1}{8}$

6. $\frac{24}{32} \div \frac{8}{32} = \frac{24 \div 8}{1} = 3$

7. $\frac{40}{50} \div \frac{5}{50} = \frac{40 \div 5}{1} = 8$

8. $\frac{60}{72} \div \frac{6}{72} = \frac{60 \div 6}{1} = 10$

9. $\frac{4}{5} \times \frac{1}{6} = \frac{4}{30} = \frac{2}{15}$

10. $\frac{5}{6} \times \frac{2}{8} = \frac{10}{48} = \frac{5}{24}$

11. $\frac{3}{5} \times \frac{4}{7} = \frac{12}{35}$

12. $\frac{2}{3} \times \frac{1}{3} = \frac{2}{9}$ sq ft

13. $\frac{3}{4} \times \frac{3}{4} = \frac{9}{16}$ sq ft

14. $\frac{1}{2} \times 2 \times \frac{5}{7} = \frac{5}{7}$ sq in

15. $\frac{5}{6} \times \frac{3}{5} = \frac{15}{30} = \frac{1}{2}$ of the job

16. $5 \times 5 \times 5 = 125$ cu ft

    $125 \times 63 = 7,875$ lb

17. $577 \div 3 = 192\frac{1}{3}$ yd

18. $10 - 5\frac{1}{2} = 9\frac{2}{2} - 5\frac{1}{2} = 4\frac{1}{2}$ years

19. Answers will vary.

20. $8 \times 7 = 56$

    $56 - 1 = 55$

    $55 \div 5 = 11$

    $11 - 3 = 8$

Lesson Practice 22A

1. done

2. done

3. $9\frac{8}{24} - 6\frac{9}{24} = 8\frac{32}{24} - 6\frac{9}{24} = 2\frac{23}{24}$

4. $10\frac{15}{18} - 3\frac{6}{18} = 7\frac{9}{18} = 7\frac{1}{2}$

5. $7\frac{27}{45} - 2\frac{40}{45} = 6\frac{72}{45} - 2\frac{40}{45} = 4\frac{32}{45}$

6. $9\frac{2}{16} - 8\frac{8}{16} = 8\frac{18}{16} - 8\frac{8}{16} = \frac{10}{16} = \frac{5}{8}$

7. $8\frac{4}{8} - 2\frac{6}{8} = 7\frac{12}{8} - 2\frac{6}{8} = 5\frac{6}{8} = 5\frac{3}{4}$

8. $5\frac{6}{18} - 2\frac{15}{18} = 4\frac{24}{18} - 2\frac{15}{18} = 2\frac{9}{18} = 2\frac{1}{2}$

9. $11\frac{4}{12} - 3\frac{3}{12} = 8\frac{1}{12}$

10. $26\frac{10}{15} - 22\frac{12}{15} = 25\frac{25}{15} - 22\frac{12}{15} = 3\frac{13}{15}$

Lesson Practice 22B

1. $10\frac{7}{28} - 3\frac{16}{28} = 9\frac{35}{28} - 3\frac{16}{28} = 6\frac{19}{28}$

2. $9\frac{6}{18} - 2\frac{15}{18} = 8\frac{24}{18} - 2\frac{15}{18} = 6\frac{9}{18} = 6\frac{1}{2}$

3. $7\frac{8}{10} - 7\frac{5}{10} = \frac{3}{10}$

4. $3\frac{5}{15} - 1\frac{12}{15} = 2\frac{20}{15} - 1\frac{12}{15} = 1\frac{8}{15}$

5. $5\frac{5}{30} - 2\frac{24}{30} = 4\frac{35}{30} - 2\frac{24}{30} = 2\frac{11}{30}$

6. $8\frac{6}{10} - 4\frac{5}{10} = 4\frac{1}{10}$

7. $15\frac{20}{28} - 12\frac{21}{28} = 14\frac{48}{28} - 12\frac{21}{28} = 2\frac{27}{28}$

8. $2\frac{4}{8} - 2\frac{2}{8} = \frac{2}{8} = \frac{1}{4}$

9. $7\frac{9}{30} - 5\frac{20}{30} = 6\frac{39}{30} - 5\frac{20}{30} = 1\frac{19}{30}$

10. $8\frac{9}{18} - 4\frac{4}{18} = 4\frac{5}{18}$

Lesson Practice 22C

1. $8\frac{10}{60} - 1\frac{18}{60} = 7\frac{70}{60} - 1\frac{18}{60} = 6\frac{52}{60} = 6\frac{13}{15}$

2. $5\frac{6}{24} - 2\frac{20}{24} = 4\frac{30}{24} - 2\frac{20}{24} = 2\frac{10}{24} = 2\frac{5}{12}$

3. $14\frac{24}{32} - 11\frac{28}{32} = 13\frac{56}{32} - 11\frac{28}{32} = 2\frac{28}{32} = 2\frac{7}{8}$

4. $7\frac{8}{20} - 3\frac{5}{20} = 4\frac{3}{20}$

5. $8\frac{3}{18} - 2\frac{12}{18} = 7\frac{21}{18} - 2\frac{12}{18} = 5\frac{9}{18} = 5\frac{1}{2}$

6. $10\frac{5}{10} - 3\frac{8}{10} = 9\frac{15}{10} - 3\frac{8}{10} = 6\frac{7}{10}$

7. $8\frac{10}{14} - 2\frac{7}{14} = 6\frac{3}{14}$

8. $4\frac{4}{8} - 1\frac{6}{8} = 3\frac{12}{8} - 1\frac{6}{8} = 2\frac{6}{8} = 2\frac{3}{4}$

9. $6\frac{12}{21} - 4\frac{14}{21} = 5\frac{33}{21} - 4\frac{14}{21} = 1\frac{19}{21}$

10. $10\frac{5}{30} - 5\frac{12}{30} = 9\frac{35}{30} - 5\frac{12}{30} = 4\frac{23}{30}$

Systematic Review 22D

1. $8\frac{8}{28} - 2\frac{21}{28} = 7\frac{36}{28} - 2\frac{21}{28} = 5\frac{15}{28}$

2. $14\frac{2}{6} - 3\frac{3}{6} = 13\frac{8}{6} - 3\frac{3}{6} = 10\frac{5}{6}$

3. $6\frac{12}{15} + 1\frac{10}{15} = 7\frac{22}{15} = 8\frac{7}{15}$

4. $4\frac{3}{5} + 1\frac{2}{5} = 5\frac{5}{5} = 6$

5. $12\frac{7}{56} + 9\frac{32}{56} = 21\frac{39}{56}$

6. $\frac{1}{2} \div \frac{1}{2} = \frac{1 \div 1}{1} = 1$

7. $\frac{1}{4} \times \frac{5}{6} = \frac{5}{24}$

8. $\frac{40}{48} \div \frac{42}{48} = \frac{40 \div 42}{1} = \frac{40}{42} = \frac{20}{21}$

9. $6\frac{2}{3} = \frac{20}{3}$

10. $27\frac{4}{9} = \frac{247}{9}$

11. $13\frac{3}{5} = \frac{68}{5}$

12. done

13. done

14. $10 \times 4 = 40$ qt

15. $11 \div 4 = 2\frac{3}{4}$ gal

16. $4\frac{3}{10} - 3\frac{9}{10} = 3\frac{13}{10} - 3\frac{9}{10} = \frac{4}{10} = \frac{2}{5}$ yd

17. $2\frac{5}{8}" \times 6\frac{1}{8}"$

(Answers that are close may be accepted.)

18. $9 - 1 = 8$
$8 \times 7 = 56$
$56 + 4 = 60$
$60 \div 6 = 10$

Systematic Review 22E

1. $10\frac{7}{56} - 4\frac{48}{56} = 9\frac{63}{56} - 4\frac{48}{56} = 5\frac{15}{56}$

2. $4\frac{4}{20} - 3\frac{15}{20} = 3\frac{24}{20} - 3\frac{15}{20} = \frac{9}{20}$

3. $2\frac{5}{50} + 6\frac{10}{50} = 8\frac{15}{50} = 8\frac{3}{10}$

4. $2\frac{21}{24} + 1\frac{16}{24} = 3\frac{37}{24} = 4\frac{13}{24}$

5. $25\frac{16}{72} + 7\frac{45}{72} = 32\frac{61}{72}$

6. $\frac{1}{5} \times \frac{1}{10} = \frac{1}{50}$

7. $\frac{10}{50} \div \frac{5}{50} = \frac{10 \div 5}{1} = 2$

8. $\frac{1}{2} \times \frac{3}{5} = \frac{3}{10}$

9. $\frac{24}{32} = \frac{24}{32}$

10. $\frac{9}{90} < \frac{20}{90}$

11. $\frac{40}{88} > \frac{33}{88}$

12. $23 \div 4 = 5\frac{3}{4}$ gal

13. $5 \times 4 = 20$ qt

14. $30 \times 3 = 90$ ft

15. $5\frac{1}{2} + 3\frac{1}{2} = 8\frac{2}{2} = 9$ gallons total
$9 - 6 = 3$ gallons left
$3 \times 4 = 12$ quarts left

16. $\frac{5}{16} \div \frac{1}{8} = \frac{40}{128} \div \frac{16}{128} =$
$\frac{40 \div 16}{1} = \frac{40}{16} = \frac{5}{2} = 2\frac{1}{2}$ pieces

17. $2 \times 2 \times 2 \times 2 \times 3$

18. $4 \times 4 = 16$

19. 1 in

20. $7 + 2 = 9$
$9 \times 7 = 63$
$63 + 2 = 65$
$65 - 61 = 4$

Systematic Review 22F

1. $12\frac{14}{18} - 1\frac{9}{18} = 11\frac{5}{18}$

2. $4\frac{4}{3} - 2\frac{2}{3} = 2\frac{2}{3}$

3. $1\frac{3}{6} + 3\frac{4}{6} = 4\frac{7}{6} = 5\frac{1}{6}$

4. $4\frac{10}{15} + 3\frac{12}{15} = 7\frac{22}{15} = 8\frac{7}{15}$

5. $6\frac{24}{32} + 9\frac{28}{32} = 15\frac{52}{32} = 15\frac{13}{8} = 16\frac{5}{8}$

6. $\frac{12}{15} \div \frac{5}{15} = \frac{12 \div 5}{1} = \frac{12}{5} = 2\frac{2}{5}$

7. $\frac{3}{5} \times \frac{3}{7} = \frac{9}{35}$

8. $\frac{1}{4} \div \frac{3}{4} = \frac{1 \div 3}{1} = \frac{1}{3}$

9. $7\frac{3}{4} = \frac{31}{4}$

10. $10\frac{1}{8} = \frac{81}{8}$

11. $45\frac{1}{2} = \frac{91}{2}$

12. $24 \div 4 = 6$ gal

13. $20 \times 4 = 80$ qt

14. $15 \div 2 = 7\frac{1}{2}$ qt

15. $315 \div 35 = 9$ ft

16. $9 \div 3 = 3$ yd

17. no

18. $15 : \underline{3}, \underline{5}, \underline{15}$
$45 : \underline{3}, \underline{5}, 9, \underline{15}, 45$
GCF = 15

19. $\frac{3}{4}$ of $24 = 18$ who want to go home
$\frac{2}{3}$ of $18 = 12$ who have a ride
$18 - 12 = 6$ who have to walk

20. $8 \times 6 = 48$
$48 - 8 = 40$
$40 \div 4 = 10$
$10 \times 3 = 30$

Lesson Practice 23A

1. done
2. $\dfrac{3}{4} \times \dfrac{4}{3} = \dfrac{12}{12} = 1$
3. $\dfrac{1}{2} \times \dfrac{2}{1} = \dfrac{2}{2} = 1$
4. done
5. done
6. $\dfrac{8}{7}$
7. done
8. done
9. $\dfrac{3}{4} \times \dfrac{8}{5} = \dfrac{24}{20} = 1\dfrac{4}{20} = 1\dfrac{1}{5}$
10. $\dfrac{24}{32} \div \dfrac{20}{32} = \dfrac{24}{20} = 1\dfrac{4}{20} = 1\dfrac{1}{5}$
11. $\dfrac{9}{10} \times \dfrac{5}{1} = \dfrac{45}{10} = \dfrac{9}{2} = 4\dfrac{1}{2}$
12. $\dfrac{45}{50} \div \dfrac{10}{50} = \dfrac{45}{10} = \dfrac{9}{2} = 4\dfrac{1}{2}$
13. done
14. done
15. $\dfrac{11}{6} \div \dfrac{13}{4} = \dfrac{11}{6} \times \dfrac{4}{13} = \dfrac{44}{78} = \dfrac{22}{39}$
16. $\dfrac{15}{8} \div \dfrac{5}{8} = \dfrac{15}{8} \times \dfrac{8}{5} = \dfrac{120}{40} = 3$
17. $\dfrac{5}{3} \div \dfrac{5}{9} = \dfrac{5}{3} \times \dfrac{9}{5} = \dfrac{45}{15} = \dfrac{9}{3} = 3$ mi
18. $\dfrac{8}{3} \div \dfrac{1}{6} = \dfrac{8}{3} \times \dfrac{6}{1} = \dfrac{48}{3} = 16$ people

Lesson Practice 23B

1. $\dfrac{1}{3} \times \dfrac{3}{1} = \dfrac{3}{3} = 1$
2. $\dfrac{5}{8} \times \dfrac{8}{5} = \dfrac{40}{40} = 1$
3. $\dfrac{3}{7} \times \dfrac{7}{3} = \dfrac{21}{21} = 1$
4. $\dfrac{2}{1}$ or 2
5. $\dfrac{1}{9}$

6. $\dfrac{4}{3}$
7. $\dfrac{4}{1} \times \dfrac{2}{1} = \dfrac{8}{1} = 8$
8. $\dfrac{8}{2} \div \dfrac{1}{2} = \dfrac{8 \div 1}{1} = 8$
9. $\dfrac{7}{10} \times \dfrac{12}{7} = \dfrac{84}{70} = 1\dfrac{14}{70} = 1\dfrac{1}{5}$
10. $\dfrac{84}{120} \div \dfrac{70}{120} = \dfrac{84}{70} = 1\dfrac{14}{70} = 1\dfrac{1}{5}$
11. $\dfrac{3}{4} \times \dfrac{8}{1} = \dfrac{24}{4} = 6$
12. $\dfrac{24}{32} \div \dfrac{4}{32} = \dfrac{24 \div 4}{1} = 6$
13. $\dfrac{38}{5} \div \dfrac{3}{2} = \dfrac{38}{5} \times \dfrac{2}{3} = \dfrac{76}{15} = 5\dfrac{1}{15}$
14. $\dfrac{13}{4} \div \dfrac{74}{7} = \dfrac{13}{4} \times \dfrac{7}{74} = \dfrac{91}{296}$
15. $\dfrac{3}{2} \div \dfrac{1}{8} = \dfrac{3}{2} \times \dfrac{8}{1} = \dfrac{24}{2} = 12$
16. $\dfrac{13}{6} \div \dfrac{1}{6} = \dfrac{13}{6} \times \dfrac{6}{1} = \dfrac{78}{6} = 13$
17. $\dfrac{6}{5} \div \dfrac{9}{10} = \dfrac{6}{5} \times \dfrac{10}{9} = \dfrac{60}{45} = \dfrac{4}{3} = 1\dfrac{1}{3}$ ft
18. $\dfrac{5}{1} \div \dfrac{5}{4} = \dfrac{5}{1} \times \dfrac{4}{5} = \dfrac{20}{5} = 4$ times

Lesson Practice 23C

1. $\dfrac{1}{6} \times \dfrac{6}{1} = \dfrac{6}{6} = 1$
2. $\dfrac{2}{3} \times \dfrac{3}{2} = \dfrac{6}{6} = 1$
3. $\dfrac{4}{5} \times \dfrac{5}{4} = \dfrac{20}{20} = 1$
4. $\dfrac{6}{5}$
5. $\dfrac{1}{13}$
6. $\dfrac{4}{9}$
7. $\dfrac{5}{7} \times \dfrac{9}{2} = \dfrac{45}{14} = 3\dfrac{3}{14}$
8. $\dfrac{45}{63} \div \dfrac{14}{63} = \dfrac{45 \div 14}{1} = \dfrac{45}{14} = 3\dfrac{3}{14}$
9. $\dfrac{7}{8} \times \dfrac{3}{1} = \dfrac{21}{8} = 2\dfrac{5}{8}$
10. $\dfrac{21}{24} \div \dfrac{8}{24} = \dfrac{21 \div 8}{1} = \dfrac{21}{8} = 2\dfrac{5}{8}$
11. $\dfrac{2}{1} \times \dfrac{3}{2} = \dfrac{6}{2} = 3$

12. $\dfrac{6}{3} \div \dfrac{2}{3} = \dfrac{6 \div 2}{1} = 3$

13. $\dfrac{8}{3} \div \dfrac{11}{8} = \dfrac{8}{3} \times \dfrac{8}{11} = \dfrac{64}{33} = 1\dfrac{31}{33}$

14. $\dfrac{16}{3} \div \dfrac{8}{3} = \dfrac{16}{3} \times \dfrac{3}{8} = \dfrac{48}{24} = 2$

15. $\dfrac{14}{5} \div \dfrac{1}{10} = \dfrac{14}{5} \times \dfrac{10}{1} = \dfrac{140}{5} = 28$

16. $\dfrac{7}{4} \div \dfrac{5}{12} = \dfrac{7}{4} \times \dfrac{12}{5} = \dfrac{84}{20} = 4\dfrac{4}{20} = 4\dfrac{1}{5}$

17. $\dfrac{5}{2} \div \dfrac{1}{2} = \dfrac{5}{2} \times \dfrac{2}{1} = \dfrac{10}{2} = 5$ sections

18. $\dfrac{25}{4} \div \dfrac{5}{1} = \dfrac{25}{4} \times \dfrac{1}{5} = \dfrac{25}{20} = 1\dfrac{5}{20} = 1\dfrac{1}{4}$ yd

Systematic Review 23D

1. $\dfrac{1}{3} \times \dfrac{5}{1} = \dfrac{5}{3} = 1\dfrac{2}{3}$

2. $\dfrac{5}{15} \div \dfrac{3}{15} = \dfrac{5 \div 3}{1} = \dfrac{5}{3} = 1\dfrac{2}{3}$

3. $\dfrac{3}{4} \times \dfrac{8}{5} = \dfrac{24}{20} = \dfrac{6}{5} = 1\dfrac{1}{5}$

4. $\dfrac{24}{32} \div \dfrac{20}{32} = \dfrac{24 \div 20}{1} = \dfrac{24}{20} = \dfrac{6}{5} = 1\dfrac{1}{5}$

5. $\dfrac{9}{4} \div \dfrac{3}{5} = \dfrac{9}{4} \times \dfrac{5}{3} = \dfrac{45}{12} = \dfrac{15}{4} = 3\dfrac{3}{4}$

6. $\dfrac{11}{6} \div \dfrac{21}{10} = \dfrac{11}{6} \times \dfrac{10}{21} = \dfrac{110}{126} = \dfrac{55}{63}$

7. $9\dfrac{5}{30} - 4\dfrac{12}{30} = 8\dfrac{35}{30} - 4\dfrac{12}{30} = 4\dfrac{23}{30}$

8. $24\dfrac{4}{3} - 16\dfrac{2}{3} = 8\dfrac{2}{3}$

9. $7\dfrac{20}{32} + 2\dfrac{8}{32} = 9\dfrac{28}{32} = 9\dfrac{7}{8}$

10. done

11. done

12. $25 \div 16 = 1\dfrac{9}{16}$ lb

13. $10 \times 16 = 160$ oz

14. $\dfrac{35}{6} \div \dfrac{7}{6} = \dfrac{35}{6} \times \dfrac{6}{7} = \dfrac{210}{42} = 5$ pieces

15. $5\dfrac{4}{8} + 4\dfrac{6}{8} = 9\dfrac{10}{8}$

    $9\dfrac{10}{8} - 6\dfrac{3}{8} = 3\dfrac{7}{8}$ lb

16. $8 \times 4 \times 2 = 64$ cu ft

17. $\dfrac{1}{2} + \dfrac{1}{2} = \dfrac{2}{2} = 1$

    $1 + \dfrac{3}{4} = 1\dfrac{3}{4}$ ft

18. $6 + 3 = 9$

    $9 - 4 = 5$

    $5 + 2 = 7$

    $7 \times 7 = 49$

Systematic Review 23E

1. $\dfrac{2}{3} \div \dfrac{5}{7} = \dfrac{2}{3} \times \dfrac{7}{5} = \dfrac{14}{15}$

2. $\dfrac{14}{21} \div \dfrac{15}{21} = \dfrac{14 \div 15}{1} = \dfrac{14}{15}$

3. $\dfrac{9}{10} \div \dfrac{1}{5} = \dfrac{9}{10} \times \dfrac{5}{1} = \dfrac{45}{10} = \dfrac{9}{2} = 4\dfrac{1}{2}$

4. $\dfrac{45}{50} \div \dfrac{10}{50} = \dfrac{45 \div 10}{1} = \dfrac{45}{10} = \dfrac{9}{2} = 4\dfrac{1}{2}$

5. $\dfrac{23}{5} \div \dfrac{17}{7} = \dfrac{23}{5} \times \dfrac{7}{17} = \dfrac{161}{85} = 1\dfrac{76}{85}$

6. $\dfrac{9}{5} \div \dfrac{4}{3} = \dfrac{9}{5} \times \dfrac{3}{4} = \dfrac{27}{20} = 1\dfrac{7}{20}$

7. $4\dfrac{6}{8} - 1\dfrac{4}{8} = 3\dfrac{2}{8} = 3\dfrac{1}{4}$

8. $13\dfrac{6}{21} - 11\dfrac{7}{21} = 12\dfrac{27}{21} - 11\dfrac{7}{21} = 1\dfrac{20}{21}$

9. $3\dfrac{1}{3} + 2\dfrac{2}{3} = 5\dfrac{3}{3} = 6$

10. $\dfrac{15}{20} + \dfrac{12}{20} = \dfrac{27}{20}$

    $\dfrac{27}{20} + \dfrac{6}{10} = \dfrac{270}{200} + \dfrac{120}{200} = \dfrac{390}{200} = \dfrac{39}{20} = 1\dfrac{19}{20}$

11. $\dfrac{7}{21} + \dfrac{15}{21} = \dfrac{22}{21}$

    $\dfrac{22}{21} + \dfrac{2}{5} = \dfrac{110}{105} + \dfrac{42}{105} = \dfrac{152}{105} = 1\dfrac{47}{105}$

12. $\dfrac{3}{6} + \dfrac{4}{6} = \dfrac{7}{6}$

    $\dfrac{7}{6} + \dfrac{4}{7} = \dfrac{49}{42} + \dfrac{24}{42} = \dfrac{73}{42} = 1\dfrac{31}{42}$

13. $64 \div 16 = 4$ lb

14. $6 \times 16 = 96$ oz

15. $13 \div 4 = 3\dfrac{1}{4}$ gal

16. $\dfrac{1}{2} \times 20 \times 5 = 50$ sq ft

17. $10\dfrac{1}{2} + 7\dfrac{1}{2} = 17\dfrac{2}{2} = 18$ ft; $18 \div 3 = 6$ yd

18. $\dfrac{93}{10} \div \dfrac{12}{5} = \dfrac{93}{10} \times \dfrac{5}{12} = \dfrac{465}{120}$

$= 3\dfrac{105}{120} = 3\dfrac{7}{8}$ ft

19. $7 \times 7 \times 7 = 343$ cu ft

20. $11 - 4 = 7$

$7 + 3 = 10$

$10 + 12 = 22$

$22 \times 2 = 44$

Systematic Review 23F

1. $\dfrac{5}{9} \div \dfrac{2}{3} = \dfrac{5}{9} \times \dfrac{3}{2} = \dfrac{15}{18} = \dfrac{5}{6}$

2. $\dfrac{15}{27} \div \dfrac{18}{27} = \dfrac{15 \div 18}{1} = \dfrac{15}{18} = \dfrac{5}{6}$

3. $\dfrac{1}{2} \div \dfrac{1}{3} = \dfrac{1}{2} \times \dfrac{3}{1} = \dfrac{3}{2} = 1\dfrac{1}{2}$

4. $\dfrac{3}{6} \div \dfrac{2}{6} = \dfrac{3 \div 2}{1} = \dfrac{3}{2} = 1\dfrac{1}{2}$

5. $\dfrac{11}{4} \div \dfrac{5}{8} = \dfrac{11}{4} \times \dfrac{8}{5} = \dfrac{88}{20} = 4\dfrac{8}{20} = 4\dfrac{2}{5}$

6. $\dfrac{12}{5} \div \dfrac{7}{6} = \dfrac{12}{5} \times \dfrac{6}{7} = \dfrac{72}{35} = 2\dfrac{2}{35}$

7. $11\dfrac{8}{7} - 6\dfrac{4}{7} = 5\dfrac{4}{7}$

8. $3\dfrac{2}{16} - 2\dfrac{8}{16} = 2\dfrac{18}{16} - 2\dfrac{8}{16} = \dfrac{10}{16} = \dfrac{5}{8}$

9. $5\dfrac{5}{50} + 2\dfrac{20}{50} = 7\dfrac{25}{50} = 7\dfrac{1}{2}$

10. $\dfrac{3}{4} = \dfrac{6}{8} = \dfrac{9}{12}$

11. $\dfrac{1}{8} = \dfrac{2}{16} = \dfrac{3}{24}$

12. $\dfrac{7}{10} = \dfrac{14}{20} = \dfrac{21}{30}$

13. $21 \div 16 = \dfrac{21}{16} = 1\dfrac{5}{16}$ lb

14. $14 \div 2 = 7$ qt

15. $25 \times 3 = 75$ ft

16. $\dfrac{15}{4} \div \dfrac{5}{1} = \dfrac{15}{4} \times \dfrac{1}{5} = \dfrac{15}{20} = \dfrac{3}{4}$ lb

17. $14 \times 11 \times 9 = 1{,}386$ cu ft

18. $\dfrac{1}{10}$

19. $\dfrac{3}{4}$ of $20 = 15$ gal used

$20 - 15 = 5$ gal left

$5 \times 4 = 20$ qt left

20. $15 \div 5 = 3$

$3 + 5 = 8$

$8 \times 6 = 48$

$48 + 2 = 50$

Lesson Practice 24A

1. done

2. done

3. $\dfrac{10}{9}$

4. $\dfrac{1}{7}$

5. done

6. done

7. $\dfrac{1}{8} \cdot 8A = 40 \cdot \dfrac{1}{8}$

$A = 5$

8. $8(5) = 40$

$40 = 40$

9. $\dfrac{1}{3} \cdot 3D = 27 \cdot \dfrac{1}{3}$

$D = 9$

10. $3(9) = 27$

$27 = 27$

11. $\dfrac{1}{11} \cdot 11Y = 121 \cdot \dfrac{1}{11}$

$Y = 11$

12. $11(11) = 121$

$121 = 121$

13. $\dfrac{1}{3} \cdot 3R = 12 \cdot \dfrac{1}{3}$

$R = 4$ rabbits

14. $\dfrac{1}{4} \cdot 4D = 48 \cdot \dfrac{1}{4}$

$D = \$12$

Lesson Practice 24B

1. $\dfrac{3}{2}$

2. $\dfrac{1}{10}$

3. $\dfrac{8}{1} = 8$

4. $\dfrac{1}{2}$

5. $\dfrac{1}{8} \cdot 8Z = 72 \cdot \dfrac{1}{8}$

   $Z = 9$

6. $8(9) = 72$

   $72 = 72$

7. $\dfrac{1}{10} \cdot 10C = 100 \cdot \dfrac{1}{10}$

   $C = 10$

8. $10(10) = 100$

   $100 = 100$

9. $\dfrac{1}{5} \cdot 5F = 25 \cdot \dfrac{1}{5}$

   $F = 5$

10. $5(5) = 25$

    $25 = 25$

11. $\dfrac{1}{2} \cdot 2A = 6 \cdot \dfrac{1}{2}$

    $A = 3$

12. $2(3) = 6$

    $6 = 6$

13. $\dfrac{1}{2} \cdot 2A = 12 \cdot \dfrac{1}{2}$

    $A = 6$ years old

14. $\dfrac{1}{5} \cdot 5B = 15 \cdot \dfrac{1}{5}$

    $B = 3$ books

Lesson Practice 24C

1. $\dfrac{9}{7}$

2. $\dfrac{1}{8}$

3. $\dfrac{2}{3}$

4. $\dfrac{1}{32}$

5. $\dfrac{1}{4} \cdot 4B = 8 \cdot \dfrac{1}{4}$

   $B = 2$

6. $4(2) = 8$

   $8 = 8$

7. $\dfrac{1}{5} \cdot 5E = 20 \dfrac{1}{5}$

   $E = 4$

8. $5(4) = 20$

   $20 = 20$

9. $\dfrac{1}{2} \cdot 2H = 24 \cdot \dfrac{1}{2}$

   $H = 12$

10. $2(12) = 24$

    $24 = 24$

11. $\dfrac{1}{3} \cdot 3C = 63 \cdot \dfrac{1}{3}$

    $C = 21$

12. $3(21) = 63$

    $63 = 63$

13. $5A = 50$

    $\dfrac{1}{5} \cdot 5A = 50 \cdot \dfrac{1}{5}$

    $A = 10$ years old

    Student may have used
    a different letter.

14. $9X = 81$

    $\dfrac{1}{9} \cdot 9X = 81 \cdot \dfrac{1}{9}$

    $X = 9$ ft

Systematic Review 24D

1. $\dfrac{1}{7} \cdot 7F = 49 \cdot \dfrac{1}{7}$

   $F = 7$

2. $7(7) = 49$

   $49 = 49$

3. $\dfrac{1}{12} \cdot 12R = 36 \cdot \dfrac{1}{12}$

   $R = 3$

4. $12(3) = 36$

   $36 = 36$

5. $\dfrac{1}{2} \div \dfrac{1}{4} = \dfrac{1}{2} \times \dfrac{4}{1} = \dfrac{4}{2} = 2$

6. $\dfrac{4}{8} \div \dfrac{2}{8} = \dfrac{4 \div 2}{1} = 2$

7. $\dfrac{2}{5} \div \dfrac{11}{6} = \dfrac{2}{5} \times \dfrac{6}{11} = \dfrac{12}{55}$

8. $\frac{11}{2} \div \frac{22}{9} = \frac{11}{2} \times \frac{9}{22} = \frac{99}{44} = \frac{9}{4} = 2\frac{1}{4}$

9. $8\frac{10}{35} - 3\frac{28}{35} = 7\frac{45}{35} - 3\frac{28}{35} = 4\frac{17}{35}$

10. $16\frac{5}{8} - 10\frac{3}{8} = 6\frac{2}{8} = 6\frac{1}{4}$

11. $5\frac{14}{20} + 4\frac{10}{20} = 9\frac{24}{20} = 10\frac{4}{20} = 10\frac{1}{5}$

12. done

13. done

14. $60 \div 12 = 5$ ft

15. $8 \times 12 = 96$ in

16. $4X = 16$

$\frac{1}{4} \cdot 4X = 16 \cdot \frac{1}{4}$

$X = 4$ gifts

17. $5 \times 16 = 80$ oz

18. $9 \times 9 = 81$

$81 - 1 = 80$

$80 \div 2 = 40$

$40 \div 5 = 8$

12. $\frac{5}{6} \times \frac{2}{3} = \frac{10}{18} = \frac{5}{9}$

13. $\frac{4}{5} \times \frac{2}{7} = \frac{8}{35}$

14. $\frac{1}{9} \times \frac{3}{8} = \frac{3}{72} = \frac{1}{24}$

15. $3S = 45$; $S = 15$ years

16. $A = \frac{5}{8} \times \frac{5}{8} = \frac{25}{64}$ sq in

$P = \frac{5}{8} + \frac{5}{8} + \frac{5}{8} + \frac{5}{8} = \frac{20}{8} = \frac{5}{2} = 2\frac{1}{2}$ in

17. $32 \div 3 = 10\frac{2}{3}$ yd

18. $32 \times 12 = 384$ in

19. $15 \times 15 = 225$

20. $30 - 10 = 20$

$20 - 2 = 18$

$18 \div 6 = 3$

$3 \times 7 = 21$

Systematic Review 24E

1. $\frac{1}{6} \cdot 6G = 42 \cdot \frac{1}{6}$

$G = 7$

2. $6(7) = 42$

$42 = 42$

3. $\frac{1}{9} \cdot 9S = 54 \cdot \frac{1}{9}$

$S = 6$

4. $9(6) = 54$

$54 = 54$

5. $\frac{6}{8} \div \frac{1}{2} = \frac{6}{8} \times \frac{2}{1} = \frac{12}{8} = \frac{3}{2} = 1\frac{1}{2}$

6. $\frac{12}{16} \div \frac{8}{16} = \frac{12 \div 8}{1} = \frac{12}{8} = \frac{3}{2} = 1\frac{1}{2}$

7. $\frac{34}{5} \div \frac{5}{3} = \frac{34}{5} \times \frac{3}{5} = \frac{102}{25} = 4\frac{2}{25}$

8. $\frac{25}{3} \div \frac{45}{8} = \frac{25}{3} \times \frac{8}{45} = \frac{200}{135} = 1\frac{65}{135} = 1\frac{13}{27}$

9. $4\frac{1}{4} + 2\frac{3}{4} = 6\frac{4}{4} = 7$

10. $2\frac{5}{50} + 6\frac{10}{50} = 8\frac{15}{50} = 8\frac{3}{10}$

11. $13\frac{9}{18} - 5\frac{14}{18} = 12\frac{27}{18} - 5\frac{14}{18} = 7\frac{13}{18}$

Systematic Review 24F

1. $\frac{1}{8} \cdot 8H = 72 \cdot \frac{1}{8}$

$H = 9$

2. $8(9) = 72$

$72 = 72$

3. $\frac{1}{11} \cdot 11T = 55 \cdot \frac{1}{11}$

$T = 5$

4. $11(5) = 55$

$55 = 55$

5. $\frac{1}{4} \div \frac{1}{3} = \frac{1}{4} \times \frac{3}{1} = \frac{3}{4}$

6. $\frac{3}{12} \div \frac{4}{12} = \frac{3 \div 4}{1} = \frac{3}{4}$

7. $\frac{7}{2} \div \frac{13}{3} = \frac{7}{2} \times \frac{3}{13} = \frac{21}{26}$

8. $\frac{23}{5} \div \frac{5}{3} = \frac{23}{5} \times \frac{3}{5} = \frac{69}{25} = 2\frac{19}{25}$

9. $8\frac{3}{4} - 8\frac{1}{4} = \frac{2}{4} = \frac{1}{2}$

10. $3\frac{14}{16} + 2\frac{8}{16} = 5\frac{22}{16} = 6\frac{6}{16} = 6\frac{3}{8}$

11. $7\frac{16}{40} - 3\frac{25}{40} = 6\frac{56}{40} - 3\frac{25}{40} = 3\frac{31}{40}$

12. $\dfrac{3}{4} \times \dfrac{1}{8} = \dfrac{3}{32}$

13. $\dfrac{1}{3} \times \dfrac{3}{7} = \dfrac{3}{21} = \dfrac{1}{7}$

14. $\dfrac{2}{5} \times \dfrac{5}{6} = \dfrac{10}{30} = \dfrac{1}{3}$

15. $6D = 18; \ D = 3$ dimes

16. $5 + 10 + 5 + 10 = 30$ ft
    $30 \div 3 = 10$ yd

17. $5\dfrac{4}{8} - 4\dfrac{6}{8} = 4\dfrac{12}{8} - 4\dfrac{6}{8} = \dfrac{6}{8} = \dfrac{3}{4}$ gal

18. $\dfrac{3}{4}$ of $12 = 9$ in

19. $\dfrac{3}{4} \times 4 = \dfrac{12}{4} = 3$ qt

20. $20 \div 4 = 5$
    $5 + 4 = 9$
    $9 \times 2 = 18$
    $18 \div 6 = 3$

Lesson Practice 25A

1. done

2. done

3. $\dfrac{1}{\cancel{6}_{2}} \times \dfrac{\cancel{3}^{1}}{8} = \dfrac{1}{16}$

4. $\dfrac{1}{\cancel{2}_{1}} \times \dfrac{\cancel{4}^{2}}{5} = \dfrac{2}{5}$

5. $\dfrac{1}{4} \times \dfrac{\cancel{7}^{1}}{11} \times \dfrac{\cancel{4}^{1}}{\cancel{7}_{1}} = \dfrac{1}{11}$

6. $\dfrac{\cancel{4}}{\cancel{5}_{1}} \times \dfrac{1}{2} \times \dfrac{\cancel{5}}{\cancel{8}_{2}} = \dfrac{1}{4}$

7. $\dfrac{1}{\cancel{5}_{1}} \times \dfrac{\cancel{5}^{1}}{6} = \dfrac{1}{6}$

8. $\dfrac{1}{\cancel{2}_{1}} \times \dfrac{\cancel{6}^{2}}{7} \times \dfrac{\cancel{2}^{1}}{\cancel{3}_{1}} = \dfrac{2}{7}$

9. $\dfrac{1}{\cancel{4}_{2}} \times \dfrac{\cancel{3}^{1}}{5} \times \dfrac{\cancel{2}^{1}}{\cancel{3}_{1}} = \dfrac{1}{10}$

10. done

11. $\dfrac{11}{\cancel{5}_{1}} \times \dfrac{\cancel{10}^{2}}{7} \times \dfrac{8}{3} = \dfrac{176}{21} = 8\dfrac{8}{21}$

12. $\dfrac{\cancel{12}^{3}}{\cancel{7}_{1}} \times \dfrac{\cancel{7}^{1}}{\cancel{4}_{1}} = \dfrac{3}{1} = 3$

13. $\dfrac{\cancel{5}^{1}}{6} \times \dfrac{\cancel{7}^{1}}{\cancel{5}_{1}} \times \dfrac{1}{\cancel{7}_{1}} = \dfrac{1}{6}$

14. $\dfrac{1}{2} \times \dfrac{1}{2} \times \dfrac{1}{2} = \dfrac{1}{8}$ of a pie

15. $\dfrac{\cancel{2}}{\cancel{3}_{1}} \times \dfrac{1}{\cancel{2}_{1}} \times \dfrac{\cancel{3}}{4} = \dfrac{1}{4}$ of the snacks

Lesson Practice 25B

1. $\dfrac{4}{\cancel{6}_{3}} \times \dfrac{\cancel{2}^{1}}{5} = \dfrac{4}{15}$

2. $\dfrac{1}{\cancel{6}_{3}} \times \dfrac{\cancel{3}^{1}}{11} \times \dfrac{\cancel{2}^{1}}{\cancel{3}_{1}} = \dfrac{1}{33}$

3. $\dfrac{\cancel{3}^{1}}{\cancel{8}_{4}} \times \dfrac{\cancel{2}^{1}}{5} \times \dfrac{3}{3} = \dfrac{3}{20}$

4. $\dfrac{4}{5} \times \dfrac{\cancel{2}^{1}}{\cancel{6}_{3}} = \dfrac{4}{15}$

5. $\dfrac{\cancel{2}^{1}}{5} \times \dfrac{\cancel{7}^{1}}{\cancel{8}_{4}} \times \dfrac{\cancel{4}^{1}}{\cancel{7}_{1}} = \dfrac{1}{5}$

6. $\dfrac{1}{\cancel{3}_{1}} \times \dfrac{\cancel{4}^{1}}{9} \times \dfrac{\cancel{3}^{1}}{\cancel{8}_{2}} = \dfrac{1}{18}$

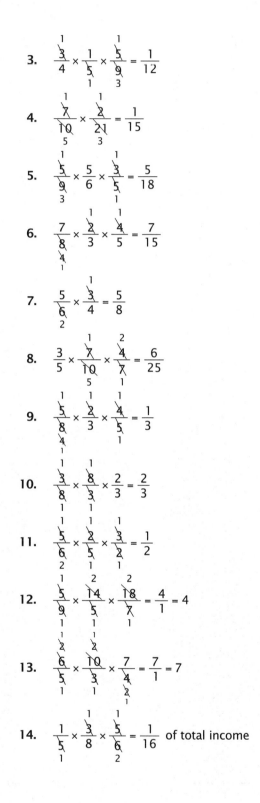

7. $\dfrac{1}{2} \times \dfrac{4}{6} = \dfrac{1}{3}$

8. $\dfrac{1}{3} \times \dfrac{6}{7} \times \dfrac{1}{5} = \dfrac{2}{35}$

9. $\dfrac{5}{8} \times \dfrac{1}{2} \times \dfrac{2}{5} = \dfrac{1}{8}$

10. $\dfrac{25}{6} \times \dfrac{8}{5} \times \dfrac{1}{10} = \dfrac{2}{3}$

11. $\dfrac{1}{2} \times \dfrac{4}{11} \times \dfrac{11}{6} = \dfrac{1}{3}$

12. $\dfrac{1}{6} \times \dfrac{7}{4} \times \dfrac{12}{7} = \dfrac{1}{2}$

13. $\dfrac{9}{2} \times \dfrac{5}{3} \times \dfrac{11}{5} = \dfrac{33}{2} = 16\dfrac{1}{2}$

14. $\dfrac{5}{2} \times \dfrac{3}{5} \times \dfrac{1}{3} = \dfrac{1}{2}$ of a bushel

15. $\dfrac{7}{10} \times \dfrac{5}{6} \times \dfrac{1}{7} = \dfrac{1}{12}$ of the profit

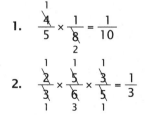

$144 \div 12 = 12$

$12 \times 1 = \$12$ for the son

Lesson Practice 25C

1. $\dfrac{4}{5} \times \dfrac{1}{8} = \dfrac{1}{10}$

2. $\dfrac{2}{3} \times \dfrac{5}{6} \times \dfrac{3}{5} = \dfrac{1}{3}$

3. $\dfrac{3}{4} \times \dfrac{1}{5} \times \dfrac{5}{9} = \dfrac{1}{12}$

4. $\dfrac{7}{10} \times \dfrac{2}{21} = \dfrac{1}{15}$

5. $\dfrac{5}{9} \times \dfrac{5}{6} \times \dfrac{3}{5} = \dfrac{5}{18}$

6. $\dfrac{7}{8} \times \dfrac{2}{3} \times \dfrac{4}{5} = \dfrac{7}{15}$

7. $\dfrac{5}{6} \times \dfrac{3}{4} = \dfrac{5}{8}$

8. $\dfrac{3}{5} \times \dfrac{7}{10} \times \dfrac{4}{7} = \dfrac{6}{25}$

9. $\dfrac{5}{8} \times \dfrac{2}{3} \times \dfrac{4}{5} = \dfrac{1}{3}$

10. $\dfrac{3}{8} \times \dfrac{8}{3} \times \dfrac{2}{3} = \dfrac{2}{3}$

11. $\dfrac{5}{6} \times \dfrac{2}{5} \times \dfrac{3}{2} = \dfrac{1}{2}$

12. $\dfrac{5}{9} \times \dfrac{14}{5} \times \dfrac{18}{7} = \dfrac{4}{1} = 4$

13. $\dfrac{6}{5} \times \dfrac{10}{3} \times \dfrac{7}{4} = \dfrac{7}{1} = 7$

14. $\dfrac{1}{5} \times \dfrac{3}{8} \times \dfrac{5}{6} = \dfrac{1}{16}$ of total income

15. $\dfrac{\cancel{9}^{\,1}_{\,3}}{2} \times \dfrac{1}{\cancel{3}_{\,1}} \times \dfrac{1}{\cancel{3}_{\,1}} = \dfrac{1}{2}$ of a pie

Systematic Review 25D

1. $\dfrac{\cancel{4}^{\,2}}{\cancel{7}} \times \dfrac{\cancel{7}}{\cancel{2}} \times \dfrac{5}{\cancel{6}_{\,3}} = \dfrac{5}{3} = 1\dfrac{2}{3}$

   When the result of simplifying is 1, it is not necessary to write it each time.

2. $\dfrac{\cancel{2}}{\cancel{9}_{\,3}} \times \dfrac{\cancel{3}}{\cancel{5}} \times \dfrac{\cancel{5}}{\cancel{2}} = \dfrac{1}{3}$

3. $\dfrac{1}{9} \cdot 9G = 36 \cdot \dfrac{1}{9}$

   $G = 4$

4. $9(4) = 36$

   $36 = 36$

5. $\dfrac{1}{12} \cdot 12V = 144 \cdot \dfrac{1}{12}$

   $V = 12$

6. $12(12) = 144$

   $144 = 144$

7. $\dfrac{27}{5} \div \dfrac{17}{5} = \dfrac{27 \div 17}{1} = \dfrac{27}{17} = 1\dfrac{10}{17}$

8. $\dfrac{82}{7} \div \dfrac{8}{7} = \dfrac{\cancel{82}^{\,41}}{\cancel{7}} \times \dfrac{\cancel{7}}{\cancel{8}_{\,4}} = \dfrac{41}{4} = 10\dfrac{1}{4}$

9. $5\dfrac{1}{4} - 2\dfrac{3}{4} = 4\dfrac{5}{4} - 2\dfrac{3}{4} = 2\dfrac{2}{4} = 2\dfrac{1}{2}$

10. $7\dfrac{20}{50} - 3\dfrac{35}{50} = 6\dfrac{70}{50} - 3\dfrac{35}{50} = 3\dfrac{35}{50} = 3\dfrac{7}{10}$

11. $9\dfrac{4}{12} + 6\dfrac{3}{12} = 15\dfrac{7}{12}$

12. done

13. done

14. $3 \times 2,000 = 6,000$ lb

15. $\dfrac{\cancel{5}}{8} \times \dfrac{1}{2} \times \dfrac{1}{\cancel{5}} = \dfrac{1}{16}$ of the pizza

16. $6X = 300$

    $\dfrac{1}{6} \cdot 6X = 300 \cdot \dfrac{1}{6}$

    $X = 50$ years old

17. $\dfrac{3}{4} \times 2,000 = \dfrac{6,000}{4}$

    $= 1,500$ lb or $\dfrac{3}{4}$ of 2,000:

    $2,000 \div 4 = 500; \; 500 \times 3 = 1,500$ lb

18. $40 - 5 = 35$

    $35 \div 7 = 5$

    $5 + 3 = 8$

    $8 \times 7 = 56$

Systematic Review 25E

1. $\dfrac{1}{\cancel{3}} \times \dfrac{\cancel{15}^{\,5}}{\cancel{8}_{\,2}} \times \dfrac{\cancel{4}}{\cancel{5}} = \dfrac{1}{2}$

2. $\dfrac{17}{\cancel{6}} \times \dfrac{\cancel{6}}{\cancel{5}} \times \dfrac{\cancel{25}^{\,5}}{8} = \dfrac{85}{8} = 10\dfrac{5}{8}$

3. $\dfrac{1}{7} \cdot 7H = 140 \cdot \dfrac{1}{7}$

   $H = 20$

4. $7(20) = 140$

   $140 = 140$

5. $\dfrac{1}{16} \cdot 16W = 48 \cdot \dfrac{1}{16}$

   $W = 3$

6. $16(3) = 48$

   $48 = 48$

7. $\dfrac{13}{4} \div \dfrac{21}{5} = \dfrac{13}{4} \times \dfrac{5}{21} = \dfrac{65}{84}$

8. $\dfrac{4}{11} \div \dfrac{2}{11} = \dfrac{4 \div 2}{1} = 2$

9. $4\dfrac{2}{12} - 1\dfrac{6}{12} = 3\dfrac{14}{12} - 1\dfrac{6}{12} = 2\dfrac{8}{12} = 2\dfrac{2}{3}$

10. $8\dfrac{4}{8} - 2\dfrac{6}{8} = 7\dfrac{12}{8} - 2\dfrac{6}{8} = 5\dfrac{6}{8} = 5\dfrac{3}{4}$

11. $7\dfrac{12}{15} + 2\dfrac{5}{15} = 9\dfrac{17}{15} = 10\dfrac{2}{15}$

12. $\dfrac{2}{3} = \dfrac{4}{6} = \dfrac{6}{9} = \dfrac{8}{12}$

13. $\dfrac{1}{10} = \dfrac{2}{20} = \dfrac{3}{30} = \dfrac{4}{40}$

14. $7 \times 2,000 = 14,000$ lb

15. $9,000 \div 2,000 = 4\frac{1000}{2000} = 4\frac{1}{2}$ tons

16. $\frac{1}{2} \times \frac{2000}{1} = \frac{2000}{2} = 1,000$ lb

17. $\frac{\cancel{10}^{2}}{\cancel{3}} \times \frac{1}{3} \times \frac{3}{\cancel{5}} = \frac{2}{3}$ acre

18. $\frac{5}{2} \times \frac{4}{1} = \frac{20}{2} = 10$ qt

19. $\frac{35}{4} \div \frac{5}{1} = \frac{\cancel{35}^{7}}{4} \times \frac{1}{\cancel{5}} = \frac{7}{4} = 1\frac{3}{4}$ yd

20. $60 + 3 = 63$
$63 \div 9 = 7$
$7 + 4 = 11$
$11 \times 3 = 33$

15. $7 \div 2 = 3\frac{1}{2}$ qt

16. $\frac{3}{\cancel{4}} \times \frac{\cancel{16}^{4}}{1} = \frac{12}{1} = 12$ oz

17. $\frac{11}{\cancel{2}} \times \frac{\cancel{12}^{6}}{1} = \frac{66}{1} = 66$ in

18. $\frac{11}{2} \div \frac{3}{1} = \frac{11}{2} \times \frac{1}{3} = \frac{11}{6} = 1\frac{5}{6}$ yd

19. $\frac{1}{3} \times \frac{1}{2} \times \frac{1}{4} = \frac{1}{24}$ of original price

20. $8 \times 6 = 48$
$48 + 2 = 50$
$50 \div 5 = 10$
$10 \times 10 = 100$

Systematic Review 25F

1. $\frac{3}{\cancel{5}} \times \frac{\cancel{5}}{4} \times \frac{1}{4} = \frac{3}{16}$

2. $\frac{\cancel{15}^{5}}{\cancel{8}} \times \frac{5}{6} \times \frac{\cancel{8}}{\cancel{3}} = \frac{25}{6} = 4\frac{1}{6}$

3. $\frac{1}{6} \cdot 6K = 54 \cdot \frac{1}{6}$
$K = 9$

4. $6(9) = 54$
$54 = 54$

5. $\frac{1}{25} \cdot 25X = 450 \cdot \frac{1}{25}$
$X = 18$

6. $25(18) = 450$

7. $\frac{16}{9} \div \frac{5}{3} = \frac{16}{\cancel{9}_{3}} \times \frac{\cancel{3}}{5} = \frac{16}{15} = 1\frac{1}{15}$

8. $\frac{7}{8} \div \frac{1}{8} = \frac{7 \div 1}{1} = 7$

9. $10\frac{4}{12} - 3\frac{9}{12} = 9\frac{16}{12} - 3\frac{9}{12} = 6\frac{7}{12}$

10. $6\frac{10}{15} - 2\frac{12}{15} = 5\frac{25}{15} - 2\frac{12}{15} = 3\frac{13}{15}$

11. $9\frac{4}{6} + 7\frac{5}{6} = 16\frac{9}{6} = 17\frac{3}{6} = 17\frac{1}{2}$

12. $\frac{4}{5} = \frac{8}{10} = \frac{12}{15} = \frac{16}{20}$

13. $\frac{3}{7} = \frac{6}{14} = \frac{9}{21} = \frac{12}{28}$

14. $4 \times 2,000 = 8,000$ lb

Lesson Practice 26A

1. done

2. done

3. $4B + 8 = 28$
$\underline{\quad -8 \quad -8}$
$4B = 20$
$\frac{1}{4} \cdot 4B = 20 \cdot \frac{1}{4}$
$B = 5$

4. $4(5) + 8 = 28$
$20 + 8 = 28$
$28 = 28$

5. $2Y - 11 = \quad 7$
$\underline{\quad +11 \quad +11}$
$2Y = \quad 18$
$\frac{1}{2} \cdot 2Y = 18 \cdot \frac{1}{2}$
$Y = 9$

6. $2(9) - 11 = 7$
$18 - 11 = 7$
$7 = 7$

7. $5C + 8 = \quad 18$
$\underline{\quad -8 \quad -8}$
$5C = 10$
$\frac{1}{5} \cdot 5C = 10 \cdot \frac{1}{5}$
$C = 2$

8. $5(2) + 8 = 18$
$10 + 8 = 18$
$18 = 18$

9. $5Z - 2 = 38$
$\underline{+2 \quad +2}$
$5Z = 40$
$\frac{1}{5} \cdot 5Z = 40 \cdot \frac{1}{5}$
$Z = 8$

10. $5(8) - 2 = 38$
$40 - 2 = 38$
$38 = 38$

11. $3D + 9 = 21$
$\underline{-9 \quad -9}$
$3D = 12$
$\frac{1}{3} \cdot 3D = 12 \cdot \frac{1}{3}$
$D = 4$

12. $3(4) + 9 = 21$
$12 + 9 = 21$
$21 = 21$

13. $7A - 7 = 42$
$\underline{+7 \quad +7}$
$7A = 49$
$\frac{1}{7} \cdot 7A = 49 \cdot \frac{1}{7}$
$A = 7$

14. $7(7) - 7 = 42$
$49 - 7 = 42$
$42 = 42$

Lesson Practice 26B

1. $5B + 3 = 18$
$\underline{-3 \quad -3}$
$5B = 15$
$\frac{1}{5} \cdot 5B = 15 \cdot \frac{1}{5}$
$B = 3$

2. $5(3) + 3 = 18$
$15 + 3 = 18$
$18 = 18$

3. $12T - 1 = 35$
$\underline{+1 \quad +1}$
$12T = 36$
$\frac{1}{12} \cdot 12T = 36 \cdot \frac{1}{12}$
$T = 3$

4. $12(3) - 1 = 35$
$36 - 1 = 35$
$35 = 35$

5. $4C + 7 = 31$
$\underline{-7 \quad -7}$
$4C = 24$
$\frac{1}{4} \cdot 4C = 24 \cdot \frac{1}{4}$
$C = 6$

6. $4(6) + 7 = 31$
$24 + 7 = 31$
$31 = 31$

7. $6S - 13 = 11$
$\underline{+13 \quad +13}$
$6S = 24$
$\frac{1}{6} \cdot 6S = 24 \cdot \frac{1}{6}$
$S = 4$

8. $6(4) - 13 = 11$
$24 - 13 = 11$
$11 = 11$

9. $10D + 11 = 91$
$\underline{-11 \quad -11}$
$10D = 80$
$\frac{1}{10} \cdot 10D = 80 \cdot \frac{1}{10}$
$D = 8$

10. $10(8) + 11 = 91$
$80 + 11 = 91$
$91 = 91$

11. $7W - 13 = 22$
$\underline{+13 \quad +13}$
$7W = 35$
$\frac{1}{7} \cdot 7W = 35 \cdot \frac{1}{7}$
$W = 5$

12. $7(5) - 13 = 22$
$35 - 13 = 22$
$22 = 22$

13. $3E + 17 = 20$
$\underline{ -17 \;\; -17}$
$3E = 3$
$\frac{1}{3} \cdot 3E = 3 \cdot \frac{1}{3}$
$E = 1$

14. $3(1) + 17 = 20$
$3 + 17 = 20$
$20 = 20$

Lesson Practice 26C

1. $5F + 13 = 38$
$\underline{ -13 \;\; -13}$
$5F = 25$
$\frac{1}{5} \cdot 5F = 25 \cdot \frac{1}{5}$
$F = 5$

2. $5(5) + 13 = 38$
$25 + 13 = 38$
$38 = 38$

3. $8V - 14 = 26$
$\underline{ +14 \;\; +14}$
$8V = 40$
$\frac{1}{8} \cdot 8V = 40 \cdot \frac{1}{8}$
$V = 5$

4. $8(5) - 14 = 26$
$40 - 14 = 26$
$26 = 26$

5. $7G + 21 = 56$
$\underline{ -21 \;\; -21}$
$7G = 35$
$\frac{1}{7} \cdot 7G = 35 \cdot \frac{1}{7}$
$G = 5$

6. $7(5) + 21 = 56$
$35 + 21 = 56$
$56 = 56$

7. $5W - 12 = 53$
$\underline{ +12 \;\; +12}$
$5W = 65$
$\frac{1}{5} \cdot 5W = 65 \cdot \frac{1}{5}$
$W = 13$

8. $5(13) - 12 = 53$
$65 - 12 = 53$
$53 = 53$

9. $6H + 7 = 25$
$\underline{ -7 \;\; -7}$
$6H = 18$
$\frac{1}{6} \cdot 6H = 18 \cdot \frac{1}{6}$
$H = 3$

10. $6(3) + 7 = 25$
$18 + 7 = 25$
$25 = 25$

11. $11X - 16 = 72$
$\underline{ +16 \;\; +16}$
$11X = 88$
$\frac{1}{11} \cdot 11X = 88 \cdot \frac{1}{11}$
$X = 8$

12. $11(8) - 16 = 72$
$88 - 16 = 72$
$72 = 72$

13. $2J + 17 = 37$
$\underline{ -17 \;\; -17}$
$2J = 20$
$\frac{1}{2} \cdot 2J = 20 \cdot \frac{1}{2}$
$J = 10$

14. $2(10) + 17 = 37$
$20 + 17 = 37$
$37 = 37$

Systematic Review 26D

1. $6Y - 12 = 12$
$6Y = 24$
$\frac{1}{6} \cdot 6 = 24 \cdot \frac{1}{6}$
$Y = 4$

2. $6(4) - 12 = 12$
$24 - 12 = 12$
$12 = 12$

3. $4K + 6 = 46$
$4K = 40$
$\frac{1}{4} \cdot 4K = 40 \cdot \frac{1}{4}$
$K = 10$

4. $4(10) + 6 = 46$
$40 + 6 = 46$
$46 = 46$

5. $\frac{\cancel{25}^5}{3} \times \frac{\cancel{34}^2}{\cancel{5}} \times \frac{1}{\cancel{17}} = \frac{10}{3} = 3\frac{1}{3}$

6. $\frac{\cancel{3}}{\cancel{11}} \times \frac{\cancel{11}}{\cancel{5}} \times \frac{\cancel{10}^2}{\cancel{3}} = \frac{2}{1} = 2$

7. $\frac{5}{16} \div \frac{2}{8} = \frac{5}{\cancel{16}_2} \times \frac{\cancel{8}}{2} = \frac{5}{4} = 1\frac{1}{4}$

8. $\frac{23}{5} \div \frac{13}{6} = \frac{23}{5} \times \frac{6}{13} = \frac{138}{65} = 2\frac{8}{65}$

9. $\frac{6}{12} + \frac{10}{12} = \frac{16}{12} = \frac{4}{3}$
$\frac{4}{3} + \frac{5}{9} = \frac{36}{27} + \frac{15}{27} = \frac{51}{27} = \frac{17}{9} = 1\frac{8}{9}$

10. $\frac{3}{12} + \frac{8}{12} = \frac{11}{12}$
$\frac{11}{12} + \frac{7}{12} = \frac{18}{12} = \frac{3}{2} = 1\frac{1}{2}$

11. $2 \times 5{,}280 = 10{,}560$ ft

12. $7 \times 5{,}280 = 36{,}960$ ft

13. $\frac{1}{2} \times 5{,}280 = 2{,}640$ ft

14. $\frac{7}{2} \times \frac{27}{4} = \frac{189}{8} = 23\frac{5}{8}$ sq ft

15. no

16. $8D = 120$
$D = 15$
$\frac{1}{8} \cdot 8D = 120 \cdot \frac{1}{8}$
$D = 15$ years old

17. $12 \times 12 = 144$

18. $16 \div 2 = 8$
$8 + 2 = 10$
$10 \times 3 = 30$
$30 \div 6 = 5$

Systematic Review 26E

1. $3Z - 13 = 20$
$3Z = 33$
$\frac{1}{3} \cdot 3Z = 33 \cdot \frac{1}{3}$
$Z = 11$

2. $3(11) - 13 = 20$
$33 - 13 = 20$
$20 = 20$

3. $8L + 7 = 55$
$8L = 48$
$L = 6$

4. $8(6) + 7 = 55$
$48 + 7 = 55$
$55 = 55$

5. $\frac{\cancel{22}^{11}}{7} \times \frac{1}{\cancel{2}} \times \frac{5}{12} = \frac{55}{84}$

6. $\frac{5}{8} \times \frac{\cancel{10}}{\cancel{3}} \times \frac{21}{\cancel{10}}^7 = \frac{35}{8} = 4\frac{3}{8}$

7. $\frac{1}{3} \div \frac{5}{18} = \frac{1}{\cancel{3}} \times \frac{\cancel{18}^6}{5} = \frac{6}{5} = 1\frac{1}{5}$

8. $\frac{16}{3} \div \frac{3}{2} = \frac{16}{3} \times \frac{2}{3} = \frac{32}{9} = 3\frac{5}{9}$

9. $\frac{9}{15} > \frac{5}{15}$

10. $\frac{16}{32} = \frac{16}{32}$

11. $\frac{7}{21} < \frac{12}{21}$

12. $5 \times 5{,}280 = 26{,}400$ ft

13. $\frac{2}{3} \times \frac{5280}{1} = \frac{3520}{1} = 3{,}520$ ft

14. $\frac{7}{2} \times 16 = 56$ oz

15. $25\frac{4}{12} + 1\frac{9}{12} = 26\frac{13}{12} = 27\frac{1}{12}$ mi

16. $25: \underline{5}, 25$
$45: 3, \underline{5}, 9, 15, 45$
GCF $= 5$

17. yes

18. $\frac{61}{10} \div \frac{3}{2} = \frac{61}{\cancel{10}_5} \times \frac{\cancel{2}}{3} = \frac{61}{15} = 4\frac{1}{15};$

5 bags

19. $9 \times 9 = 81$

20. $35 \div 7 = 5$

$5 + 4 = 9$

$9 \times 6 = 54$

$54 + 1 = 55$

Systematic Review 26F

1. $7A + 8 = 29$

$7A = 21$

$\frac{1}{7} \cdot 7A = 21 \cdot \frac{1}{7}$

$A = 3$

2. $7(3) + 8 = 29$

$21 + 8 = 29$

$29 = 29$

3. $9M - 9 = 54$

$9M = 63$

$\frac{1}{9} \cdot 9M = 63 \cdot \frac{1}{9}$

$M = 7$

4. $9(7) - 9 = 54$

$63 - 9 = 54$

$54 = 54$

5. $\frac{\cancel{6}}{5} \times \frac{3}{2} \times \frac{5}{\cancel{6}} = \frac{3}{2} = 1\frac{1}{2}$

6. $\frac{3}{\cancel{7}} \times \frac{5}{6} \times \frac{\cancel{7}}{3} = \frac{5}{6}$

7. $\frac{5}{8} \div \frac{1}{12} = \frac{5}{\cancel{8}} \times \frac{\cancel{12}}{1} = \frac{15}{2} = 7\frac{1}{2}$

8. $\frac{12}{5} \div \frac{36}{25} = \frac{\cancel{12}}{5} \times \frac{\cancel{25}}{\cancel{36}} = \frac{5}{3} = 1\frac{2}{3}$

9. $\frac{3}{4} - \frac{1}{4} = \frac{2}{4} = \frac{1}{2}$

10. $\frac{5}{10} - \frac{2}{10} = \frac{3}{10}$

11. $\frac{24}{30} - \frac{5}{30} = \frac{19}{30}$

12. $35 \div 3 = 11\frac{2}{3}$ yd

13. $8 \times 2 = 16$ pt

14. $\frac{3}{2} \times 4 = 6$ qt

15. $10\frac{4}{8} - 6\frac{6}{8} = 9\frac{12}{8} - 6\frac{6}{8} = 3\frac{6}{8} = 3\frac{3}{4}$ tons

$\frac{15}{\cancel{4}} \times \frac{\overset{500}{\cancel{2000}}}{1} = 7{,}500$ lb

16. 3x3x7

17. $\frac{13}{\cancel{2}} \times \overset{6}{\cancel{12}} = 78$ in

18. $\frac{7}{8} \times \overset{660}{\cancel{5{,}280}} = 4{,}620$ ft

$4{,}620 \div 3 = 1{,}540$ yd

19. $23 \times 23 = 529$

20. $42 \div 6 = 7$

$7 - 1 = 6$

$6 \times 6 = 36$

$36 + 4 = 40$

Lesson Practice 27A

1. done

2. $\frac{22}{7}\left(3^2\right) = \frac{22}{7} \times \frac{9}{1} = \frac{198}{7} = 28\frac{2}{7}$ sq ft

3. $\frac{22}{7}\left(21^2\right) = \frac{22}{\cancel{7}} \times \frac{\overset{63}{\cancel{441}}}{1}$

$= \frac{1386}{1} = 1{,}386$ sq in

4. $\frac{22}{7}\left(6^2\right) = \frac{22}{7} \times \frac{36}{1} = \frac{792}{7} = 113\frac{1}{7}$ sq ft

5. $\frac{22}{7}\left(4^2\right) = \frac{22}{7} \times \frac{16}{1} = \frac{352}{7} = 50\frac{2}{7}$ sq yd

6. done

7. $\frac{2}{1} \times \frac{22}{7} \times \frac{3}{1} = \frac{132}{7} = 18\frac{6}{7}$ ft

8. $\frac{2}{1} \times \frac{22}{\cancel{7}} \times \frac{\overset{3}{\cancel{21}}}{1} = \frac{132}{1} = 132$ in

9. $\frac{2}{1} \times \frac{22}{7} \times \frac{6}{1} = \frac{264}{7} = 37\frac{5}{7}$ ft

10. $\frac{2}{1} \times \frac{22}{\cancel{7}} \times \frac{\cancel{7}}{1} = \frac{44}{1} = 44$ in

Lesson Practice 27B

1. $\frac{22}{7}(4^2) = \frac{22}{7} \times \frac{16}{1} = \frac{352}{7} = 50\frac{2}{7}$ sq in

2. $\frac{22}{7}(28^2) = \frac{22}{\cancel{7}} \times \frac{\cancel{784}^{112}}{1}$
   $= \frac{2464}{1} = 2,464$ sq ft

3. $\frac{22}{7}(11^2) = \frac{22}{7} \times \frac{121}{1}$
   $= \frac{2662}{7} = 380\frac{2}{7}$ sq in

4. $\frac{22}{7}(70^2) = \frac{22}{\cancel{7}} \times \frac{\cancel{4900}^{700}}{1}$
   $= \frac{15400}{1} = 15,400$ sq ft

5. $\frac{22}{7}(10^2) = \frac{22}{7} \times \frac{100}{1}$
   $= \frac{2200}{7} = 314\frac{2}{7}$ sq ft

6. $\frac{2}{1} \times \frac{22}{7} \times \frac{4}{1} = \frac{176}{7} = 25\frac{1}{7}$ in

7. $\frac{2}{1} \times \frac{22}{\cancel{7}} \times \frac{\cancel{28}^4}{1} = \frac{176}{1} = 176$ ft

8. $\frac{2}{1} \times \frac{22}{7} \times \frac{11}{1} = \frac{484}{7} = 69\frac{1}{7}$ in

9. $\frac{2}{1} \times \frac{22}{\cancel{7}} \times \frac{\cancel{70}^{10}}{1} = \frac{440}{1} = 440$ ft

10. $\frac{2}{1} \times \frac{22}{7} \times \frac{9}{1} = \frac{396}{7} = 56\frac{4}{7}$ in

Lesson Practice 27C

1. $\frac{22}{7}(49^2) = \frac{22}{\cancel{7}} \times \frac{\cancel{2401}^{343}}{1}$
   $= \frac{7546}{1} = 7,546$ sq in

2. $\frac{22}{7}(5^2) = \frac{22}{7} \times \frac{25}{1} = \frac{550}{7} = 78\frac{4}{7}$ sq ft

3. $\frac{22}{7}(8^2) = \frac{22}{7} \times \frac{64}{1} = \frac{1408}{7} = 201\frac{1}{7}$ sq in

4. $\frac{22}{7}(42^2) = \frac{22}{\cancel{7}} \times \frac{\cancel{1764}^{252}}{1}$
   $= \frac{5544}{1} = 5,544$ sq ft

5. $\frac{22}{7}(12^2) = \frac{22}{7} \times \frac{144}{1} = \frac{3168}{7}$
   $= 452\frac{4}{7}$ sq ft

6. $\frac{2}{1} \times \frac{22}{\cancel{7}} \times \frac{\cancel{49}^7}{1} = \frac{308}{1} = 308$ in

7. $\frac{2}{1} \times \frac{22}{7} \times \frac{5}{1} = \frac{220}{7} = 31\frac{3}{7}$ ft

8. $\frac{2}{1} \times \frac{22}{7} \times \frac{8}{1} = \frac{352}{7} = 50\frac{2}{7}$ in

9. $\frac{2}{1} \times \frac{22}{\cancel{7}} \times \frac{\cancel{42}^6}{1} = \frac{264}{1} = 264$ ft

10. $\frac{2}{1} \times \frac{22}{\cancel{7}} \times \frac{\cancel{35}^5}{1} = \frac{220}{1} = 220$ mi

Systematic Review 27D

1. $\frac{22}{7}(7^2) = \frac{22}{\cancel{7}} \times \frac{\cancel{49}^7}{1} = \frac{154}{1} = 154$ sq in

2. $\frac{2}{1} \times \frac{22}{\cancel{7}} \times \frac{\cancel{7}}{1} = \frac{44}{1} = 44$ in

3. $\frac{22}{7}(9^2) = \frac{22}{7} \times \frac{81}{1} = \frac{1782}{7} = 254\frac{4}{7}$ sq ft

4. $\frac{2}{1} \times \frac{22}{7} \times \frac{9}{1} = \frac{396}{7} = 56\frac{4}{7}$ ft

5. $3P - 15 = 12$
   $3P = 27$
   $\frac{1}{3} \cdot 3P = 27 \cdot \frac{1}{3}$
   $P = 9$

6. $3(9) - 15 = 12$
   $27 - 15 = 12$
   $12 = 12$

7. $\frac{\cancel{6}}{\cancel{5}} \times \frac{\cancel{5}}{\cancel{12}_2} \times \frac{3}{7} = \frac{3}{14}$

8. $\frac{3}{\cancel{5}} \times \frac{\cancel{20}^4}{\cancel{14}_2} \times \frac{\cancel{7}}{2} = \frac{12}{4} = 3$

9. $\dfrac{9}{11} \div \dfrac{1}{11} = \dfrac{9 \div 1}{1} = 9$

10. $\dfrac{9}{4} \div \dfrac{27}{22} = \dfrac{9}{4} \times \dfrac{22}{27} = \dfrac{11}{6} = 1\dfrac{5}{6}$

11. done

12. 8 pt = 8 pt

13. 8 oz > 7 oz

14. $\dfrac{22}{7}\left(16^2\right) = \dfrac{22}{7} \times \dfrac{256}{1}$

$\quad\quad = \dfrac{5632}{7} = 804\dfrac{4}{7}$ sq in

15. $\dfrac{1}{2} \times \dfrac{1}{2} \times \dfrac{1}{2} = \dfrac{1}{8}$ cu ft

16. $\dfrac{1}{2} \times 12 = 6$ in; $6 \times 6 \times 6 = 216$ cu in

17. $2\dfrac{1}{4} + 5\dfrac{1}{4} = 7\dfrac{2}{4} = 7\dfrac{1}{2}$

$\quad\quad 7\dfrac{1}{2} + 1\dfrac{1}{2} = 8\dfrac{2}{2} = 9$ in

18. $10 \times 10 = 100$

$\quad\quad 100 \div 2 = 50$

$\quad\quad 50 + 3 = 53$

$\quad\quad 53 - 4 = 49$

Systematic Review 27E

1. $\dfrac{22}{7}\left(56^2\right) = \dfrac{22}{7} \times \dfrac{3136}{1} = 9856$ sq in

2. $\dfrac{2}{1} \times \dfrac{22}{7} \times \dfrac{56}{1} = \dfrac{352}{1} = 352$ in

3. $\dfrac{22}{7}\left(6^2\right) = \dfrac{22}{7} \times \dfrac{36}{1} = \dfrac{792}{7} = 113\dfrac{1}{7}$ sq ft

4. $\dfrac{2}{1} \times \dfrac{22}{7} \times \dfrac{6}{1} = \dfrac{264}{7} = 37\dfrac{5}{7}$ ft

5. $9D + 14 = 23$

$\quad\quad 9D = 9$

$\quad\quad \dfrac{1}{9} \cdot 9D = 9 \cdot \dfrac{1}{9}$

$\quad\quad D = 1$

6. $9(1) + 14 = 23$

$\quad\quad 9 + 14 = 23$

$\quad\quad 23 = 23$

7. $\dfrac{9}{4} \times \dfrac{8}{3} \times \dfrac{2}{9} = \dfrac{4}{3} = 1\dfrac{1}{3}$

8. $\dfrac{5}{6} \times \dfrac{9}{2} \times \dfrac{4}{3} = \dfrac{5}{1} = 5$

9. $\dfrac{4}{7} \div \dfrac{7}{3} = \dfrac{4}{7} \times \dfrac{3}{7} = \dfrac{12}{49}$

10. $\dfrac{32}{3} \div \dfrac{17}{4} = \dfrac{32}{3} \times \dfrac{4}{17} = \dfrac{128}{51} = 2\dfrac{26}{51}$

11. 15 qt < 32 qt

12. 4,000 lb > 1,500 lb

13. 1,760 ft = 1,760 ft

14. $\dfrac{3}{4} \times 2,000 = 1,500$ lb

15. $\dfrac{22}{7}\left(14^2\right) = \dfrac{22}{7} \times \dfrac{196}{1} = \dfrac{616}{1} = 616$ sq ft

16. $\dfrac{2}{1} \times \dfrac{22}{7} \times \dfrac{14}{1} = \dfrac{88}{1} = 88$ ft

17. $\dfrac{1}{2} \times \dfrac{9}{2} \times \dfrac{8}{1} = \dfrac{1}{2} \times \dfrac{9}{1} \times \dfrac{4}{1}$

$\quad\quad = \dfrac{18}{1} = 18$ sq in

18. $\dfrac{21}{2} \div \dfrac{7}{6} = \dfrac{21}{2} \times \dfrac{6}{7} = \dfrac{9}{1} = 9$ cakes

19. $\quad 3A = 90$

$\quad\quad \dfrac{1}{3} \cdot 3A = 90 \cdot \dfrac{1}{3}$

$\quad\quad A = 30$ years old

20. $11 \times 3 = 33$

$\quad\quad 33 - 1 = 32$

$\quad\quad 32 \div 8 = 4$

$\quad\quad 4 \times 7 = 28$

Systematic Review 27F

1. $\dfrac{22}{7}\left(21^2\right) = \dfrac{22}{7} \times \dfrac{441}{1}$

$\quad\quad = \dfrac{1386}{1} = 1,386$ sq in

2. $\frac{2}{1} \times \frac{22}{7} \times \frac{\overset{3}{\cancel{21}}}{1} = \frac{132}{1} = 132$ in

3. $\frac{22}{7}\left(10^2\right) = \frac{22}{7} \times \frac{100}{1}$

$= \frac{2200}{7} = 314\frac{2}{7}$ sq ft

4. $\frac{2}{1} \times \frac{22}{7} \times \frac{10}{1} = \frac{440}{7} = 62\frac{6}{7}$ ft

5. $4Q - 5 = 3$
   $\quad\; +5 \; +5$
   $4Q \quad = 8$
   $\frac{1}{4} \cdot 4Q = \frac{1}{4} \cdot 8$
   $\quad\; Q = 2$

6. $4(2) - 5 = 3$
   $8 - 5 = 3$
   $3 = 3$

7. $\frac{\overset{3}{\cancel{12}}}{5} \times \frac{11}{\cancel{4}} = \frac{33}{5} = 6\frac{3}{5}$

8. $\frac{\overset{9}{\cancel{45}}}{\cancel{8}} \times \frac{\cancel{8}}{\cancel{5}} \times \frac{15}{7} = \frac{135}{7} = 19\frac{2}{7}$

9. $\frac{4}{10} \div \frac{2}{5} = \frac{\cancel{4}}{\underset{\cancel{2}}{\cancel{10}}} \times \frac{\cancel{5}}{\cancel{2}} = \frac{1}{1} = 1$

10. $\frac{66}{7} \div \frac{21}{10} = \frac{66}{7} \times \frac{10}{\underset{7}{\cancel{21}}}^{\;22} = \frac{220}{49} = 4\frac{24}{49}$

11. 204 in > 17 in

12. 10,000 lb = 10,000 lb

13. 3,960 ft < 4,000 ft

14. $\frac{22}{7}\left(1^2\right) = \frac{22}{7} \times \frac{1}{1} = \frac{22}{7} = 3\frac{1}{7}$ sq mi

15. $\frac{22}{7}\left(2^2\right) = \frac{22}{7} \times \frac{4}{1} = \frac{88}{7} = 12\frac{4}{7}$ sq mi; no

16. $10 - 6\frac{3}{4} = 3\frac{1}{4}$ lb

17. $\frac{1}{2} \times \frac{9}{1} \times \frac{10}{3} = \frac{15}{1} = 15$ sq in

18. $20\frac{1}{3} + 20\frac{1}{3} = 40\frac{2}{3}$
    $30\frac{1}{2} + 30\frac{1}{2} = 60\frac{2}{2} = 61$
    $40\frac{2}{3} + 61 = 101\frac{2}{3}$ ft

19. $101\frac{2}{3} \times \frac{3}{5} = \frac{\overset{61}{\cancel{305}}}{\cancel{3}} \times \frac{3}{5} = 61$ ft

20. $12 + 8 = 20$
    $20 - 10 = 10$
    $10 \times 4 = 40$
    $40 + 7 = 47$

Lesson Practice 28A

1. done

2. done
   Feel free to use canceling to simplify your work.

3. $\frac{5}{1} \cdot \frac{1}{5}A = \frac{3}{1} \cdot \frac{5}{1}$
   $A = \frac{15}{1} = 15$

4. $\frac{1}{5}(15) = 3$
   $3 = 3$

5. $\frac{1}{7}Y - 4 = 2$
   $\frac{1}{7}Y = 6$
   $\frac{7}{1} \cdot \frac{1}{7}Y = \frac{6}{1} \cdot \frac{7}{1}$
   $Y = \frac{42}{1} = 42$

6. $\frac{1}{7}(42) - 4 = 2$
   $6 - 4 = 2$
   $2 = 2$

7. $\frac{3}{8}B + 8 = 14$
   $\frac{3}{8}B = 6$
   $\frac{8}{3} \cdot \frac{3}{8}B = \frac{6}{1} \cdot \frac{8}{3}$
   $B = \frac{48}{3} = 16$

8. $\frac{3}{8} \cdot \frac{16}{1} + 8 = 14$

$\frac{48}{8} + 8 = 14$

$6 + 8 = 14$

$14 = 14$

9. $\frac{3}{2} \cdot \frac{2}{3} Z = \frac{12}{1} \cdot \frac{3}{2}$

$Z = \frac{36}{2} = 18$

10. $\frac{2}{3} \cdot \frac{18}{1} = 12$

$\frac{36}{3} = 12$

$12 = 12$

11. $\frac{2}{5} C - 2 = 2$

$\frac{2}{5} C = 4$

$\frac{5}{2} \cdot \frac{2}{5} C = \frac{4}{1} \cdot \frac{5}{2}$

$C = \frac{20}{2} = 10$

12. $\frac{2}{5} \cdot \frac{10}{1} - 2 = 2$

$\frac{20}{5} - 2 = 2$

$4 - 2 = 2$

$2 = 2$

Lesson Practice 28B

1. $\frac{1}{4} G - 4 = 2$

$\frac{1}{4} G = 6$

$\frac{4}{1} \cdot \frac{1}{4} G = \frac{6}{1} \cdot \frac{4}{1}$

$G = \frac{24}{1} = 24$

2. $\frac{1}{4}(24) - 4 = 2$

$\frac{1}{4} \cdot \frac{24}{1} - 4 = 2$

$\frac{24}{4} - 4 = 2$

$6 - 4 = 2$

$2 = 2$

3. $\frac{5}{4} \cdot \frac{4}{5} D = \frac{40}{1} \cdot \frac{5}{4}$

$D = \frac{200}{4} = 50$

4. $\frac{4}{5} \cdot \frac{50}{1} = 40$

$\frac{200}{5} = 40$

$40 = 40$

5. $\frac{1}{10} H - 4 = 0$

$\frac{1}{10} H = 4$

$\frac{10}{1} \cdot \frac{1}{10} H = \frac{4}{1} \cdot \frac{10}{1}$

$H = \frac{40}{1} = 40$

6. $\frac{1}{10}(40) - 4 = 0$

$4 - 4 = 0$

$0 = 0$

7. $\frac{5}{8} E - 5 = 20$

$\frac{5}{8} E = 25$

$\frac{8}{5} \cdot \frac{5}{8} E = \frac{25}{1} \cdot \frac{8}{5}$

$E = \frac{200}{5} = 40$

8. $\frac{5}{8}(40) - 5 = 20$

$\frac{5}{8} \cdot \frac{40}{1} - 5 = 20$

$\frac{200}{8} - 5 = 20$

$25 - 5 = 20$

9. $\frac{4}{1} \cdot \frac{1}{4} J = \frac{7}{1} \cdot \frac{4}{1}$

$J = \frac{28}{1} = 28$

10. $\frac{1}{4}(28) = 7$

$7 = 7$

11. $\frac{5}{9}F + 9 = 14$

$\frac{5}{9}F = 5$

$\frac{9}{5} \cdot \frac{5}{9}F = \frac{5}{1} \cdot \frac{9}{5}$

$F = \frac{45}{5} = 9$

12. $\frac{5}{9} \cdot \frac{45}{5} + 9 = 14$

$\frac{45}{9} + 9 = 14$

$5 + 9 = 14$

$14 = 14$

Lesson Practice 28C

1. $\frac{1}{3}K + 3 = 11$

$\frac{1}{3}K = 8$

$\frac{3}{1} \cdot \frac{1}{3}K = \frac{8}{1} \cdot \frac{3}{1}$

$K = \frac{24}{1} = 24$

2. $\frac{1}{3} \cdot \frac{24}{1} + 3 = 11$

$\frac{24}{3} + 3 = 11$

$8 + 3 = 11$

$11 = 11$

3. $\frac{5}{1} \cdot \frac{1}{5}G = \frac{8}{1} \cdot \frac{5}{1}$

$G = \frac{40}{1} = 40$

4. $\frac{1}{5} \cdot \frac{40}{1} = 8$

$\frac{40}{5} = 8$

$8 = 8$

5. $\frac{3}{5}L + 3 = 18$

$\frac{3}{5}L = 15$

$\frac{5}{3} \cdot \frac{3}{5}L = \frac{15}{1} \cdot \frac{5}{3}$

$L = \frac{75}{3} = 25$

6. $\frac{3}{5}(25) + 3 = 18$

$\frac{3}{5} \cdot \frac{25}{1} + 3 = 18$

$\frac{75}{5} + 3 = 18$

$15 + 3 = 18$

7. $\frac{5}{12}R + 5 = 25$

$\frac{5}{12}R = 20$

$\frac{12}{5} \cdot \frac{5}{12}R = \frac{20}{1} \cdot \frac{12}{5}$

$R = \frac{240}{5} = 48$

8. $\frac{5}{12} \cdot \frac{48}{1} + 5 = 25$

$\frac{240}{12} + 5 = 25$

$20 + 5 = 25$

$25 = 25$

9. $\frac{3}{2} \cdot \frac{2}{3}M = \frac{10}{1} \cdot \frac{3}{2}$

$M = \frac{30}{2} = 15$

10. $\frac{2}{3} \cdot \frac{15}{1} = 10$

$\frac{30}{3} = 10$

$10 = 10$

11. $\frac{2}{9}S + 6 = 14$

$\frac{2}{9}S = 8$

$\frac{9}{2} \cdot \frac{2}{9}S = \frac{8}{1} \cdot \frac{9}{2}$

$S = \frac{72}{2} = 36$

12. $\frac{2}{9}(36) + 6 = 14$

$\frac{2}{9} \cdot \frac{36}{1} + 6 = 14$

$\frac{72}{9} + 6 = 14$

$8 + 6 = 14$

$14 = 14$

Systematic Review 28D

1. $\frac{1}{5}T + 4 = 8$

$$\frac{1}{5}T = 4$$

$$\frac{5}{1} \cdot \frac{1}{5}T = \frac{4}{1} \cdot \frac{5}{1}$$

$$T = \frac{20}{1} = 20$$

2. $\frac{1}{5}(20) + 4 = 8$

$$\frac{1}{5} \cdot \frac{20}{1} + 4 = 8$$

$$\frac{20}{5} + 4 = 8$$

$$4 + 4 = 8$$

$$8 = 8$$

3. $\frac{3}{4}N - 6 = 21$

$$\frac{3}{4}N = 27$$

$$\frac{4}{3} \cdot \frac{3}{4}N = \frac{27}{1} \cdot \frac{4}{3}$$

$$N = \frac{108}{3}$$

$$N = 36$$

4. $\frac{3}{4}(36) - 6 = 21$

$$\frac{3}{4} \cdot \frac{36}{1} - 6 = 21$$

$$\frac{108}{4} - 6 = 21$$

$$27 - 6 = 21$$

$$21 = 21$$

5. $\frac{22}{7}(14^2) = \frac{22}{\cancel{7}} \cdot \frac{\cancel{196}^{28}}{1} = \frac{616}{1} = 616 \text{ sq ft}$

6. $\frac{2}{1} \cdot \frac{22}{\cancel{7}} \cdot \frac{\cancel{14}^{2}}{1} = \frac{88}{1} = 88 \text{ ft}$

7. $\frac{1}{\cancel{2}} \times \frac{\cancel{2}}{\cancel{3}} \times \frac{\cancel{9}^{3}}{11} = \frac{3}{11}$

8. $\frac{\cancel{7}}{8} \times \frac{8}{\cancel{3}} \times \frac{\cancel{6}}{\cancel{7}}^{2} = \frac{2}{1} = 2$

9. $\frac{3}{1} \div \frac{1}{2} = \frac{3}{1} \times \frac{2}{1} = \frac{6}{1} = 6$

10. $\frac{31}{5} \div \frac{31}{20} = \frac{\cancel{31}}{\cancel{5}} \times \frac{\cancel{20}^{4}}{\cancel{31}} = \frac{4}{1} = 4$

11. 21 ft < 36 ft

12. 10 pt < 11 pt

13. 5 qt > 4 qt

14. $2 + 7 + 9 = 18$

$18 \div 3 = 6$

15. $5 + 5 + 9 + 13 = 32$

$32 \div 4 = 8$

16. $2 + 5 + 8 + 9 = 24$

$24 \div 4 = 6$

17. $77 + 80 + 95 + 100 = 352$

$352 \div 4 = 88$ average score

18. $2 + 7 = 9$

$9 \times 7 = 63$

$63 + 2 = 65$

$65 + 5 = 70$

Systematic Review 28E

1. $\frac{1}{6}U + 2 = 3$

$$\frac{1}{6}U = 1$$

$$\frac{6}{1} \cdot \frac{1}{6}U = \frac{1}{1} \cdot \frac{6}{1}$$

$$U = \frac{6}{1} = 6$$

2. $\frac{1}{6}(6) + 2 = 3$

$$\frac{1}{6} \cdot \frac{6}{1} + 2 = 3$$

$$1 + 2 = 3$$

$$3 = 3$$

3. $\frac{5}{2} \cdot \frac{2}{5}P = \frac{2}{1} \cdot \frac{5}{2}$

$$P = \frac{10}{2} = 5$$

4. $\frac{2}{5} \cdot \frac{5}{1} = 2$

$$\frac{10}{5} = 2$$

$$2 = 2$$

5. $\frac{22}{7}\left(42^2\right) = \frac{22}{\cancel{7}} \cdot \frac{\overset{252}{\cancel{1764}}}{1}$

$= \frac{5544}{1} = 5{,}544$ sq in

6. $\frac{2}{1} \times \frac{22}{\cancel{7}} \times \frac{\overset{6}{\cancel{42}}}{1} = 264$ in

7. $\frac{\overset{3}{\cancel{15}}}{\underset{2}{\cancel{14}}} \times \frac{\cancel{7}}{4} \times \frac{1}{\underset{5}{\cancel{25}}} = \frac{3}{40}$

8. $\frac{\overset{2}{\cancel{4}}}{\underset{3}{\cancel{9}}} \times \frac{5}{\cancel{2}} \times \frac{\cancel{3}}{\cancel{5}} = \frac{2}{3}$

9. $\frac{3}{6} \div \frac{2}{3} = \frac{3}{6} \times \frac{3}{2} = \frac{9}{12} = \frac{3}{4}$

10. $\frac{24}{7} \div \frac{18}{7} = \frac{24 \div 18}{1} = \frac{24}{18} = \frac{4}{3} = 1\frac{1}{3}$

11. 32 oz = 32 oz

12. 14,000 lb > 8,000 lb

13. 120 in < 125 in

14. $4 + 5 + 6 = 15$

$15 \div 3 = 5$

15. $6 + 7 + 9 + 10 = 32$

$32 \div 4 = 8$

16. $4 + 7 + 10 = 21$

$21 \div 3 = 7$

17. $4\frac{1}{2} + 4\frac{1}{2} = 8\frac{2}{2} = 9$

$6\frac{1}{3} + 6\frac{1}{3} = 12\frac{2}{3}$

$12\frac{2}{3} + 9 = 21\frac{2}{3}$ in

18. $\frac{\overset{3}{\cancel{9}}}{2} \times \frac{19}{\cancel{3}} = \frac{57}{2} = 28\frac{1}{2}$ sq in

19. $2 \times 2 \times 2 \times 2 \times 2 \times 2$

20. $3 \times 2 = 6$

$6 \times 3 = 18$

$18 + 2 = 20$

$20 - 4 = 16$

Systematic Review 28F

1. $\frac{3}{16}V - 3 = 12$

$\frac{3}{16}V = 15$

$\frac{16}{3} \cdot \frac{3}{16}V = \frac{15}{1} \cdot \frac{16}{3}$

$V = \frac{240}{3} = 80$

2. $\frac{3}{16}(80) - 3 = 12$

$\frac{3}{16} \cdot \frac{80}{1} - 3 = 12$

$\frac{240}{16} - 3 = 12$

$15 - 3 = 12$

$12 = 12$

3. $5Q + 4 = 19$

$5Q = 15$

$\frac{1}{5} \cdot 5Q = \frac{15}{1} \cdot \frac{1}{5}$

$Q = \frac{15}{5} = 3$

4. $5(3) + 4 = 19$

$15 + 4 = 19$

$19 = 19$

5. $\frac{22}{7}\left(2^2\right) = \frac{22}{7} \cdot \frac{4}{1} = \frac{88}{7} = 12\frac{4}{7}$ sq in

6. $\frac{2}{1} \cdot \frac{22}{7} \cdot \frac{2}{1} = \frac{88}{7} = 12\frac{4}{7}$ in

7. $\frac{\cancel{2}}{1} \times \frac{3}{\underset{5}{\cancel{10}}} \times \frac{5}{\underset{2}{\cancel{6}}} = \frac{1}{2}$

8. $\frac{\overset{2}{\cancel{10}}}{\underset{3}{\cancel{21}}} \times \frac{\overset{2}{\cancel{14}}}{\cancel{5}} \times \frac{\overset{2}{\cancel{6}}}{11} = \frac{8}{11}$

9. $\frac{1}{2} \div \frac{1}{10} = \frac{1}{2} \times \frac{10}{1} = \frac{10}{2} = 5$

10. $\frac{37}{8} \div \frac{41}{12} = \frac{37}{\underset{2}{\cancel{8}}} \times \frac{\overset{3}{\cancel{12}}}{41} = \frac{111}{82} = 1\frac{29}{82}$

11. 256 oz > 1 oz

12. 2,640 ft = 2,640 ft

13. 9 in < 10 in

14. $4 + 8 + 10 + 14 = 36$

$36 \div 4 = 9$

15. $6 + 7 + 9 + 14 = 36$
    $36 \div 4 = 9$

16. $6 + 6 + 12 = 24$
    $24 \div 3 = 8$

17. $2\frac{1}{4} + 1\frac{1}{4} = 3\frac{2}{4} = 3\frac{1}{2}$
    $3\frac{1}{2} + 1\frac{1}{2} = 4\frac{2}{2} = 5$
    $5 \div 3 = \frac{5}{3} = 1\frac{2}{3}$ mi

18. $\frac{4}{3} \times \frac{4}{3} \times \frac{4}{3} = \frac{64}{27} = 2\frac{10}{27}$ cu in

19. $30 : 2, \underline{3}, \underline{5}, 6, 10, \underline{15}, 30$
    $45 : \underline{3}, \underline{5}, 9, \underline{15}, 45$
    GCF $= 15$

20. $20 \div 5 = 4$
    $4 \times 3 = 12$
    $12 + 7 = 19$
    $19 - 4 = 15$

Lesson Practice 29A

1. done

2. $\frac{1}{2} = \frac{5}{10} = 0.5 = \frac{50}{100} = 0.50$

3. $\frac{4}{5} = \frac{8}{10} = 0.8 = \frac{80}{100} = 0.80$

4. done

5. $\frac{5}{5} = \frac{100}{100} = 100\%$

6. done

7. $\frac{1}{2} = \frac{50}{100} = 0.50 = 50\%$

Lesson Practice 29B

1. done

2. $\frac{3}{4} = \frac{75}{100} = 0.75$

3. $\frac{2}{5} = \frac{4}{10} = 0.4 = \frac{40}{100} = 0.40$

4. $\frac{3}{5} = \frac{60}{100} = 60\%$

5. $\frac{4}{5} = \frac{80}{100} = 80\%$

6. $\frac{1}{4} = \frac{25}{100} = 0.25 = 25\%$

7. $\frac{3}{4} = \frac{75}{100} = 0.75 = 75\%$

Lesson Practice 29C

1. $\frac{1}{5} = \frac{2}{10} = 0.2 = \frac{20}{100} = 0.20$

2. $\frac{3}{5} = \frac{6}{10} = 0.6 = \frac{60}{100} = 0.60$

3. $\frac{1}{4} = \frac{25}{100} = 0.25$

4. $\frac{5}{5} = \frac{100}{100} = 100\%$

5. $\frac{2}{5} = \frac{40}{100} = 40\%$

6. $\frac{3}{4} = \frac{75}{100} = 0.75 = 75\%$

7. $\frac{1}{2} = \frac{50}{100} = 0.50 = 50\%$

Systematic Review 29D

1. $\frac{1}{5} = \frac{20}{100} = 20\%$

2. $\frac{3}{5} = \frac{60}{100} = 60\%$

3. $\frac{1}{4} = \frac{25}{100} = 0.25 = 25\%$

4. $\frac{4}{5} = \frac{80}{100} = 0.80 = 80\%$

5. $\frac{1}{2}T + 5 = 8$
   $\frac{1}{2}T = 3$
   $\frac{2}{1} \cdot \frac{1}{2}T = \frac{3}{1} \cdot \frac{2}{1}$
   $T = \frac{6}{1} = 6$

6. $\frac{1}{2}(6) + 5 = 8$
   $3 + 5 = 8$
   $8 = 8$

7. $\frac{\cancel{3}}{\cancel{4}} \times \frac{\cancel{2}}{7} \times \frac{1}{\cancel{3}} = \frac{1}{14}$

8. $\frac{\cancel{42}}{\cancel{8}} \times \frac{47}{\cancel{6}} \times \frac{\cancel{12}}{\cancel{7}} = \frac{141}{2} = 70\frac{1}{2}$

9. 6

10. 13

11. 9

12. 31

13. XX

14. VIII

15. XVI

16. XXXIV

17. $16 + 18 + 22 = 56$

$56 \div 3 = 18\frac{2}{3}$ is average age

18. $A = \frac{22}{7}(70^2) = \frac{22}{\cancel{7}} \cdot \frac{\cancel{4900}^{700}}{1}$

$= \frac{15400}{1} = 15,400$ sq ft

$C = \frac{2}{1} \cdot \frac{22}{\cancel{7}} \cdot \frac{\cancel{70}^{10}}{1} = \frac{440}{1} = 440$ ft

11. 27

12. 19

13. VII

14. XVIII

15. XXI

16. XXXV

17. $\left(\frac{1}{2}\right)(24)\left(\frac{13}{2}\right) = 78$ sq in

18. $\frac{3}{8} \div \frac{5}{8} = \frac{3}{5}$ mi

19. $3\frac{2}{6} + 6\frac{3}{6} + 5\frac{1}{6} = 14\frac{6}{6} = 15$

$15 \div 3 = 5$ in

20. $\frac{2}{1} \cdot \frac{22}{\cancel{7}} \cdot \frac{\cancel{7}}{1} = \frac{44}{1} = 44$ ft

Systematic Review 29E

1. $\frac{2}{5} = \frac{40}{100} = 40\%$

2. $\frac{5}{5} = \frac{100}{100} = 100\%$

3. $\frac{3}{4} = \frac{75}{100} = 0.75 = 75\%$

4. $\frac{1}{2} = \frac{50}{100} = 0.50 = 50\%$

5. $\frac{4}{5}U - 11 = 33$

$\frac{4}{5}U = 44$

$\frac{5}{4} \cdot \frac{4}{5}U = \frac{44}{1} \cdot \frac{5}{4}$

$U = \frac{220}{4} = 55$

6. $\frac{4}{5}(55) - 11 = 33$

$\frac{4}{5} \cdot \frac{55}{1} - 11 = 33$

$\frac{220}{5} - 11 = 33$

$44 - 11 = 33$

$33 = 33$

7. $\frac{9}{12} + \frac{8}{12} + \frac{1}{12} = \frac{18}{12} = \frac{3}{2} = 1\frac{1}{2}$

8. $\frac{5}{15} + \frac{6}{15} + \frac{1}{15} = \frac{12}{15} = \frac{4}{5}$

9. 2

10. 17

Systematic Review 29F

1. $\frac{3}{5} = \frac{60}{100} = 60\%$

2. $\frac{1}{5} = \frac{20}{100} = 20\%$

3. $\frac{4}{5} = \frac{80}{100} = 0.80 = 80\%$

4. $\frac{1}{4} = \frac{25}{100} = 0.25 = 25\%$

5. $7X - 9 = 19$

$7X = 28$

$\frac{1}{7} \cdot 7X = 28 \cdot \frac{1}{7}$

$X = 4$

6. $7(4) - 9 = 19$

$28 - 9 = 19$

$19 = 19$

7. $16\frac{5}{9} - 8\frac{1}{9} = 8\frac{4}{9}$

8. $6\frac{5}{5} - 3\frac{4}{5} = 3\frac{1}{5}$

9. $4\frac{15}{40} + 2\frac{24}{40} = 6\frac{39}{40}$

10. 3

11. 11

12. 25

13. 39

14. V

15. XII

16. XXXVIII

17. XIV

18. $45 + 34 + 40 = 119$

 $119 \div 3 = 39\frac{2}{3}$ average miles

19. $\frac{1}{2} \times \frac{1}{4} \times \frac{4}{5} = \frac{1}{10}$ of the job

20. $\frac{22}{7}(28^2) = \frac{22}{7} \cdot \frac{\overset{112}{\cancel{784}}}{1}$

 $= \frac{2464}{1} = 2,464$ sq ft

 $2,464 > 200$; no

Lesson Practice 30A

1. done

2. done

3. $\frac{3}{5}X - \frac{1}{2} = 19\frac{1}{2}$

 $\frac{3}{5}X = 19\frac{1}{2} + \frac{1}{2}$

 $\frac{3}{5}X = 20$

 $\frac{5}{3} \cdot \frac{3}{5}X = \frac{20}{1} \cdot \frac{5}{3}$

 $X = \frac{100}{3} = 33\frac{1}{3}$

4. $\frac{3}{5} \cdot \frac{100}{3} - \frac{1}{2} = 19\frac{1}{2}$

 $\frac{300}{15} - \frac{1}{2} = 19\frac{1}{2}$

 $20 - \frac{1}{2} = 19\frac{1}{2}$

 $19\frac{1}{2} = 19\frac{1}{2}$

5. $\frac{1}{6}L + \frac{1}{3} = \frac{2}{3}$

 $\frac{1}{6}L = \frac{1}{3}$

 $\frac{6}{1} \cdot \frac{1}{6}L = \frac{1}{3} \cdot \frac{6}{1}$

 $L = \frac{6}{3} = 2$

6. $\frac{1}{6}(2) + \frac{1}{3} = \frac{2}{3}$

 $\frac{1}{6} \cdot \frac{2}{1} + \frac{1}{3} = \frac{2}{3}$

 $\frac{2}{6} + \frac{1}{3} = \frac{2}{3}$

 $\frac{1}{3} + \frac{1}{3} = \frac{2}{3}$

 $\frac{2}{3} = \frac{2}{3}$

7. $\frac{1}{2}Y + \frac{3}{4} = 7\frac{3}{4}$

 $\frac{1}{2}Y = 7$

 $\frac{2}{1} \cdot \frac{1}{2}Y = \frac{7}{1} \cdot \frac{2}{1}$

 $Y = \frac{14}{1} = 14$

8. $\frac{1}{2}(14) + \frac{3}{4} = 7\frac{3}{4}$

 $7 + \frac{3}{4} = 7\frac{3}{4}$

 $7\frac{3}{4} = 7\frac{3}{4}$

9. $\frac{4}{5}M - \frac{1}{10} = \frac{6}{10}$

 $\frac{4}{5}M = \frac{7}{10}$

 $\frac{5}{4} \cdot \frac{4}{5}M = \frac{7}{\underset{2}{\cancel{10}}} \cdot \frac{\cancel{5}}{4}$

 $M = \frac{7}{8}$

10. $\frac{4}{5} \cdot \frac{7}{8} - \frac{1}{10} = \frac{3}{5}$

 $\frac{28}{40} - \frac{1}{10} = \frac{3}{5}$

 $\frac{7}{10} - \frac{1}{10} = \frac{3}{5}$

 $\frac{6}{10} = \frac{3}{5}$

11. $\frac{1}{4}Z - \frac{2}{3} = 3\frac{1}{3}$

 $\frac{1}{4}Z = 3\frac{3}{3} = 4$

 $\frac{4}{1} \cdot \frac{1}{4}Z = 4 \cdot \frac{4}{1}$

 $Z = 16$

12. $\frac{1}{4}(16) - \frac{2}{3} = 3\frac{1}{3}$

$4 - \frac{2}{3} = 3\frac{1}{3}$

$3\frac{1}{3} = 3\frac{1}{3}$

Lesson Practice 30B

1. $\frac{1}{2}P - \frac{1}{8} = \frac{1}{4}$

$\frac{1}{2}P = \frac{3}{8}$

$\frac{2}{1} \cdot \frac{1}{2}P = \frac{3}{8} \cdot \frac{2}{1}$

$P = \frac{6}{8} = \frac{3}{4}$

2. $\frac{1}{2} \cdot \frac{3}{4} - \frac{1}{8} = \frac{1}{4}$

$\frac{3}{8} - \frac{1}{8} = \frac{1}{4}$

$\frac{1}{4} = \frac{1}{4}$

3. $\frac{1}{3}B - \frac{2}{7} = 1\frac{5}{7}$

$\frac{1}{3}B = 2$

$\frac{3}{1} \cdot \frac{1}{3}B = 2 \cdot \frac{3}{1}$

$B = 6$

4. $\frac{1}{3}(6) - \frac{2}{7} = 1\frac{5}{7}$

$2 - \frac{2}{7} = 1\frac{5}{7}$

$1\frac{5}{7} = 1\frac{5}{7}$

5. $\frac{3}{8}N + \frac{2}{8} = \frac{5}{8}$

$\frac{3}{8}N = \frac{3}{8}$

$\frac{8}{3} \cdot \frac{3}{8}N = \frac{3}{8} \cdot \frac{8}{3}$

$N = 1$

6. $\frac{3}{8}(1) + \frac{1}{4} = \frac{5}{8}$

$\frac{3}{8} + \frac{2}{8} = \frac{5}{8}$

$\frac{5}{8} = \frac{5}{8}$

7. $\frac{4}{5}A + \frac{1}{9} = 4\frac{1}{9}$

$\frac{4}{5}A = 4$

$\frac{5}{4} \cdot \frac{4}{5}A = \frac{4}{1} \cdot \frac{5}{4}$

$A = \frac{20}{4} = 5$

8. $\frac{4}{5}(5) + \frac{1}{9} = 4\frac{1}{9}$

$4 + \frac{1}{9} = 4\frac{1}{9}$

$4\frac{1}{9} = 4\frac{1}{9}$

9. $\frac{3}{4}R - \frac{4}{8} = \frac{1}{8}$

$\frac{3}{4}R = \frac{5}{8}$

$\frac{4}{3} \cdot \frac{3}{4}R = \frac{5}{8} \cdot \frac{4}{3}$

$R = \frac{20}{24} = \frac{5}{6}$

10. $\frac{3}{4} \cdot \frac{5}{6} - \frac{1}{2} = \frac{1}{8}$

$\frac{15}{24} - \frac{4}{8} = \frac{1}{8}$

$\frac{5}{8} - \frac{4}{8} = \frac{1}{8}$

$\frac{1}{8} = \frac{1}{8}$

11. $\frac{1}{6}D - \frac{1}{6} = 8\frac{5}{6}$

$\frac{1}{6}D = 9$

$\frac{6}{1} \cdot \frac{1}{6}D = 9 \cdot \frac{6}{1}$

$D = 54$

12. $\frac{1}{6}(54) - \frac{1}{6} = 8\frac{5}{6}$

$9 - \frac{1}{6} = 8\frac{5}{6}$

$8\frac{5}{6} = 8\frac{5}{6}$

Lesson Practice 30C

1. $\frac{5}{6}T - \frac{1}{3} = \frac{1}{3}$

    $\frac{5}{6}T = \frac{2}{3}$

    $\frac{6}{5} \cdot \frac{5}{6}T = \frac{2}{3} \cdot \frac{6}{5}$

    $T = \frac{12}{15} = \frac{4}{5}$

2. $\frac{5}{6} \cdot \frac{4}{5} - \frac{1}{3} = \frac{1}{3}$

    $\frac{20}{30} - \frac{1}{3} = \frac{1}{3}$

    $\frac{2}{3} - \frac{1}{3} = \frac{1}{3}$

    $\frac{1}{3} = \frac{1}{3}$

3. $\frac{3}{4}F - \frac{3}{5} = 17\frac{2}{5}$

    $\frac{3}{4}F = 18$

    $\frac{4}{3} \cdot \frac{3}{4}F = 18 \cdot \frac{4}{3}$

    $F = \frac{72}{3} = 24$

4. $\frac{3}{4}(24) - \frac{3}{5} = 17\frac{2}{5}$

    $18 - \frac{3}{5} = 17\frac{2}{5}$

    $17\frac{2}{5} = 17\frac{2}{5}$

5. $\frac{2}{3}Q + \frac{1}{6} = \frac{1}{3}$

    $\frac{2}{3}Q = \frac{1}{6}$

    $\frac{3}{2} \cdot \frac{2}{3}Q = \frac{1}{6} \cdot \frac{3}{2}$

    $Q = \frac{3}{12} = \frac{1}{4}$

6. $\frac{2}{3} \cdot \frac{1}{4} + \frac{1}{6} = \frac{1}{3}$

    $\frac{1}{6} + \frac{1}{6} = \frac{1}{3}$

    $\frac{1}{3} = \frac{1}{3}$

7. $\frac{3}{4}C + \frac{1}{3} = 15\frac{1}{3}$

    $\frac{3}{4}C = 15$

    $\frac{4}{3} \cdot \frac{3}{4}C = 15 \cdot \frac{4}{3}$

    $C = \frac{60}{3} = 20$

8. $\frac{3}{4}(20) + \frac{1}{3} = 15\frac{1}{3}$

    $15 + \frac{1}{3} = 15\frac{1}{3}$

    $15\frac{1}{3} = 15\frac{1}{3}$

9. $\frac{3}{5}S + \frac{7}{10} = \frac{9}{10}$

    $\frac{3}{5}S = \frac{2}{10}$

    $\frac{5}{3} \cdot \frac{3}{5}S = \frac{2}{10} \cdot \frac{5}{3}$

    $S = \frac{10}{30} = \frac{1}{3}$

10. $\frac{3}{5} \cdot \frac{1}{3} + \frac{7}{10} = \frac{9}{10}$

    $\frac{1}{5} + \frac{7}{10} = \frac{9}{10}$

    $\frac{2}{10} + \frac{7}{10} = \frac{9}{10}$

    $\frac{9}{10} = \frac{9}{10}$

11. $\frac{1}{6}H - \frac{1}{3} = 1\frac{2}{3}$

    $\frac{1}{6}H = 2$

    $\frac{6}{1} \cdot \frac{1}{6}H = 2 \cdot \frac{6}{1}$

    $H = 12$

12. $\frac{1}{6}(12) - \frac{1}{3} = 1\frac{2}{3}$

    $2 - \frac{1}{3} = 1\frac{2}{3}$

    $1\frac{2}{3} = 1\frac{2}{3}$

Systematic Review 30D

1. $\frac{3}{5}G + \frac{1}{2} = 9\frac{1}{2}$

   $\frac{3}{5}G = 9$

   $\frac{5}{3} \cdot \frac{3}{5}G = 9 \cdot \frac{5}{3}$

   $G = \frac{45}{3} = 15$

2. $\frac{3}{5}(15) + \frac{1}{2} = 9\frac{1}{2}$

   $9 + \frac{1}{2} = 9\frac{1}{2}$

   $9\frac{1}{2} = 9\frac{1}{2}$

3. $\frac{4}{7}U + \frac{2}{7} = \frac{6}{7}$

   $\frac{4}{7}U = \frac{4}{7}$

   $\frac{7}{4} \cdot \frac{4}{7}U = \frac{4}{7} \cdot \frac{7}{4}$

   $U = 1$

4. $\frac{4}{7}(1) + \frac{2}{7} = \frac{6}{7}$

   $\frac{6}{7} = \frac{6}{7}$

5. $\frac{2}{5} = \frac{40}{100} = 0.40 = 40\%$

6. $\frac{3}{4} = \frac{75}{100} = 0.75 = 75\%$

7. $\frac{15}{24} < \frac{16}{24}$

8. $\frac{10}{50} = \frac{10}{50}$

9. $\frac{48}{64} > \frac{36}{64}$

10. done

11. 230

12. 525

13. 1,610

14. XLIX

15. CCCLII

16. DLXXXIII

17. MMDLV

18. $\frac{7}{2} \times \frac{5}{2} \times \frac{3}{2} = \frac{105}{8} = 13\frac{1}{8}$ cu ft

Systematic Review 30E

1. $\frac{1}{3}E + \frac{5}{8} = 6\frac{5}{8}$

   $\frac{1}{3}E = 6$

   $\frac{3}{1} \cdot \frac{1}{3}E = 6 \cdot \frac{3}{1}$

   $E = 18$

2. $\frac{1}{3}(18) + \frac{5}{8} = 6\frac{5}{8}$

   $6 + \frac{5}{8} = 6\frac{5}{8}$

   $6\frac{5}{8} = 6\frac{5}{8}$

3. $\frac{1}{8}V - \frac{1}{8} = \frac{3}{4}$

   $\frac{1}{8}V = \frac{7}{8}$

   $\frac{8}{1} \cdot \frac{1}{8}V = \frac{7}{8} \cdot \frac{8}{1}$

   $V = \frac{56}{8} = 7$

4. $\frac{1}{8}(7) - \frac{1}{8} = \frac{3}{4}$

   $\frac{7}{8} - \frac{1}{8} = \frac{3}{4}$

   $\frac{6}{8} = \frac{3}{4}$

   $\frac{3}{4} = \frac{3}{4}$

5. $\frac{1}{4} = \frac{25}{100} = 0.25 = 25\%$

6. $\frac{4}{5} = \frac{80}{100} = 0.80 = 80\%$

7. $\frac{2}{3} = \frac{4}{6} = \frac{6}{9} = \frac{8}{12}$

8. $\frac{5}{8} = \frac{10}{16} = \frac{15}{24} = \frac{20}{32}$

9. 86

10. 152

11. 3,500

12. LXXIV

13. CCXI

14. MDXXII

15. 3 ft

16. 2 pt

17. 4 qt

18. 5x5 = 25

19. $4X + 16 = 64$

$\qquad 4X - 48$

$\qquad \dfrac{1}{4} \cdot 4X = 48 \cdot \dfrac{1}{4}$

$\qquad X = 12$

20. $5\dfrac{3}{8} + 5\dfrac{3}{8} + 5\dfrac{3}{8} + 5\dfrac{3}{8} = 20\dfrac{12}{8}$

$\qquad\qquad = 21\dfrac{4}{8} = 21\dfrac{1}{2}$ in

Systematic Review 30F

1. $\dfrac{5}{6}J + \dfrac{2}{5} = 10\dfrac{2}{5}$

$\qquad \dfrac{5}{6}J = 10$

$\qquad \dfrac{6}{5} \cdot \dfrac{5}{6}J = 10 \cdot \dfrac{6}{5}$

$\qquad J = \dfrac{60}{5} = 12$

2. $\dfrac{5}{6}(12) + \dfrac{2}{5} = 10\dfrac{2}{5}$

$\qquad \dfrac{5}{6} \cdot \dfrac{12}{1} + \dfrac{2}{5} = 10\dfrac{2}{5}$

$\qquad \dfrac{60}{6} + \dfrac{2}{5} = 10\dfrac{2}{5}$

$\qquad 10\dfrac{2}{5} = 10\dfrac{2}{5}$

3. $\dfrac{3}{5} = \dfrac{60}{100} = 0.60 = 60\%$

4. $\dfrac{1}{2} = \dfrac{50}{100} = 0.50 = 50\%$

5. $\dfrac{32}{15} \div \dfrac{4}{3} = \dfrac{\overset{8}{\cancel{32}}}{\underset{5}{\cancel{15}}} \times \dfrac{\cancel{3}}{\cancel{4}} = \dfrac{8}{5} = 1\dfrac{3}{5}$

6. $\dfrac{\cancel{12}}{\cancel{5}} \times \dfrac{\cancel{5}}{\cancel{4}} \times \dfrac{\overset{2}{\cancel{8}}}{7} = \dfrac{24}{7} = 3\dfrac{3}{7}$

7. $7\dfrac{5}{10} - 3\dfrac{8}{10} =$

$\qquad 6\dfrac{15}{10} - 3\dfrac{8}{10} = 3\dfrac{7}{10}$

8. $3\dfrac{15}{40} + 1\dfrac{32}{40} = 4\dfrac{47}{40} = 5\dfrac{7}{40}$

9. 91

10. 410

11. 1,135

12. LVIII

13. CDXII

14. MMCCLX

15. 16 oz

16. 12 in

17. 5,280 ft

18. 2,000 lb

19. $\dfrac{22}{7}\left(35^2\right) = \dfrac{22}{\cancel{7}} \cdot \dfrac{\overset{175}{\cancel{1225}}}{1}$

$\qquad\qquad = \dfrac{3850}{1} = 3,850$ sq mi

20. $\dfrac{2}{1} \cdot \dfrac{22}{\cancel{7}} \cdot \dfrac{\overset{5}{\cancel{35}}}{1} = 220$ mi

Appendix Lesson A1

1. done

2. $9 + 11 = 20$

$\qquad 20 \div 2 = 10$

$\qquad 10 \times 6 = 60$ sq in

3. $6 + 10 = 16$

$\qquad 16 \div 2 = 8$

$\qquad 8 \times 4 = 32$ sq ft

4. $6 + 12 = 18$

$\qquad 18 \div 2 = 9$

$\qquad 9 \times 7 = 63$ sq ft

5. $3 + 5 = 8$

$\qquad 8 \div 2 = 4$

$\qquad 4 \times 2 = 8$ sq ft

6. $8 + 10 = 18$

$\qquad 18 \div 2 = 9$

$\qquad 9 \times 5 = 45$ sq in

7. $7 + 9 = 16$

$\qquad 16 \div 2 = 8$

$\qquad 8 \times 5 = 40$ sq in

8. $1 + 3 = 4$

$\qquad 4 \div 2 = 2$

$\qquad 2 \times 2 = 4$ sq mi

Appendix Lesson A2

1. $2 + 4 = 6$
    $6 \div 2 = 3$
    $3 \times 3 = 9$ sq in
2. $3 + 17 = 20$
    $20 \div 2 = 10$
    $10 \times 6 = 60$ sq ft
3. $5 + 9 = 14$
    $14 \div 2 = 7$
    $7 \times 5 = 35$ sq ft
4. $4 + 10 = 14$
    $14 \div 2 = 7$
    $7 \times 3 = 21$ sq in
5. $8 + 12 = 20$
    $20 \div 2 = 10$
    $10 \times 11 = 110$ sq ft
6. $7 + 11 = 18$
    $18 \div 2 = 9$
    $9 \times 4 = 36$ sq in
7. $2 + 6 = 8$
    $8 \div 2 = 4$
    $4 \times 6 = 24$ sq ft
    24 plants
8. $5 + 7 = 12$
    $12 \div 2 = 6$
    $6 \times 6 = 36$ sq in

Application and Enrichment Solutions

Application and Enrichment 1G

Across

1. drama
4. estimation
6. build
8. story
9. math
10. drawing

Down

2. altogether
3. real
5. invent
7. Key

1. done
2. 6 – 4 = 2 pies
3. $25 + $42 = $67
4. Key words: "[fraction] of the cookies" indicates multiplying a fraction by a number (12 cookies). "Altogether" indicates addition.
   1/3 × 12 = 4 cookies
   1/4 × 12 = 3 cookies
   4 + 3 = 7 cookies
5. Key words: "were left" indicates subtraction; "one half of them" indicates multiplying a fraction (1/2) by a number (the four remaining trees).
   7 – 3 = 4 trees
   1/2 × 4 = 2 apple trees
6. Key words: "total" indicates addition; "3/4 of the total" indicates multiplying a fraction by a number; "have left" indicates subtraction.
   $20 + $16 = $36
   3/4 × $36 = $27
   $36 – $27 = $9

Application and Enrichment 2G

1. Colored parts should match the numerator of each fraction.

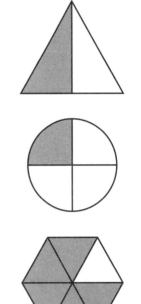

2.

3.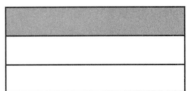

4.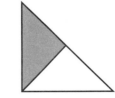

5.

6.

7. 4 eggs are broken.
   6 eggs are left.

8. 2 boxes will be donated.
6 boxes will be left.

9. 8 pieces of pie were eaten.
4 pieces were left.

Application and Enrichment 3G

1. two times
$2 \times (6 + 4) = 2 \times 10 = 20$

2. eight times
$8 \times (5 + 2) = 8 \times 7 = 56$

3. three times
$3 \times (9 - 4) = 3 \times 5 = 15$

4. ten times
$10 \times (1,268 - 345) = 10 \times 923$
$= 9,230$

5. $1/6 + 3/6 = 4/6$ of the seashells
$4/6$ of 12
$12 \div 6 = 2$
$2 \times 4 = 8$ seashells

6. $1/4 + 1/4 = 2/4$ of the prizes
$2/4$ of 8
$8 \div 4 = 2$
$2 \times 2 = 4$ prizes

7. $2/5 + 3/5 = 5/5$ of his problems
$5/5$ of 20
$20 \div 5 = 4$
$4 \times 5 = 20$ problems
Steve has solved all of the problems.

Application and Enrichment 4G

1.

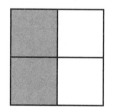

2. No, there is still the same amount of cake in the pan.
$2/4$

3. Ian's share is 1/4 of the whole cake.
Jamie's share is 1/4 of the whole cake.

4.

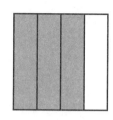

5.

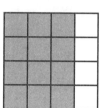

12/16 of the cake is in the pan.

6. 12/16 cut into 4 parts = 3/16

7.

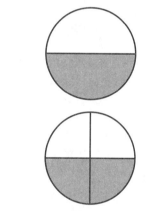

8.

2/4 of the pie is left to be eaten.

9. Each of the boys will eat 1/4 of the whole pie.
Each of the boys will eat 1/2 of the leftover pie.

10. $4/5 = 2/5 + 2/5$
$4/5 = 1/5 + 3/5$
$4/5 = 1/5 + 1/5 + 1/5 + 1/5$
Since this is addition, the order of the fractions is not important.

11. 3/6 = 1/6 + 2/6
    3/6 = 1/6 + 1/6 + 1/6
12. 7/8 = 1/8 + 6/8
    7/8 = 2/8 + 5/8
    7/8 = 3/8 + 4/8
    7/8 = 2/8 + 2/8 + 2/8 + 1/8
    7/8 = 1/8 + 1/8 + 1/8 + 1/8 + 1/8
        + 1/8 + 1/8
    There are other options as well.

Application and Enrichment 5G

1. Key words: another, in all
   Add
2. Key words: is left
   Subtract
3. Key words: difference
   Subtract
4. Key words: none
   Add
5. Key words: is left
   Subtract

6.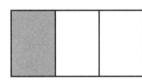
   one mile

   1/4 + 1/4 = 2/4 of a mile
7.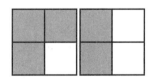

   1 − 1/3 = 3/3 − 1/3 = 2/3 of the
   garden still needs to be planted.
8.

   3/4 − 1/2 =
   6/8 − 4/8 = 2/8 or 1/4 lb

9.

   1/3 + 1/6 = 6/18 + 3/18 = 9/18 or
   1/2 of the jelly beans
10. 1/2 of 12
    12 ÷ 2 = 6
    1 × 6 = 6 jelly beans
    Counting the shaded or colored
    jelly beans in the drawing yields
    the same answer.

Application and Enrichment 6G

1.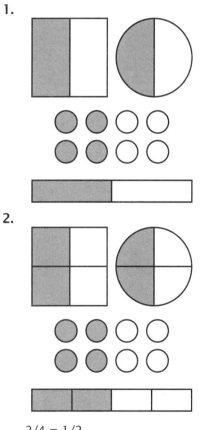

2.

   2/4 = 1/2

3.

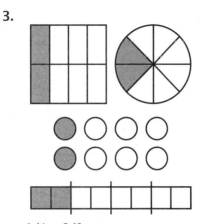

1/4 = 2/8

4. done
5. done
6. 3 × (2) = 6
7. Answers will vary.
8. 5 × (6) = 30
9. Answers will vary.
10. (3) × (6) = 18
11. Answers will vary.

Application and Enrichment 7G

Answer should be "READ FOR MEANING."

1. addition
2. addition
3. subtraction
4. addition
5. subtraction
6. subtraction
7. addition
8. subtraction
9. addition
10. addition

Application and Enrichment 8G

Note: Depending on the method the student uses to add the fractions, the final answer may differ. As long as the numerator and denominator are the same in the final answer (e.g., 4/4, 8/8, 64/64), the answer is correct.

1. 1/4 + 1/4 + 1/8 + 1/8 + 1/4 =
   1/8 + 1/8 + 1/4 + 1/4 + 1/4 =
   2/8 + 3/4 =
   8/32 + 24/32 =
   32/32 = 1
2. 1/4 + 1/8 + 1/8 + 1/2 =
   1/8 + 1/8 + 1/4 + 1/2 =
   2/8 + 1/4 + 1/2 =
   8/32 + 8/32 + 1/2 =
   16/32 + 1/2 =
   32/64 + 32/64 =
   64/64 = 1
3. 1/4 + 1/2 + 1/8 + 1/8 =
   1/8 + 1/8 + 1/4 + 1/2 =
   2/8 + 1/4 + 1/2 =
   2/8 + 2/8 + 4/8 =
   8/8 = 1

Note that, after rearranging the fractions, problem 3 was the same as problem 2. We used a different method to solve in problem 3. Since both methods produced answers equivalent to 1, both methods are correct and may be used with either problem.

Application and Enrichment 9G

1. What the recipe calls for:

What Mary needs:

1/3 × 1/3 = 1/9 of a cup
2. 2 × 1/3 = 2/3 of a cup
3. Juncos divided into 4 groups of 3 each; 3 groups = 9 juncos;
   3/4 × 12 = 9 juncos

4. 3/1 × 1/4 = 3/4 of a pizza

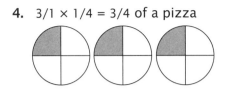

Application and Enrichment 10G

1. done
2. done
3. 1/7 of what was left over;
   what was left over = 7/8
   1/7 × 7/8 = 7/56 of a whole pie
4. 1/2 of the days last month;
   the days last month = 30
   1/2 × 30 = 15 days
5. 2/3 of what Paul did;
   what Paul did = 1/5
   2/3 × 1/5 = 2/15 of the job
6. 6 × 1/2 = 3 miles

Application and Enrichment 11G

1. 3/1 ÷ 1/6 = 18/6 ÷ 1/6 = 18
   times. (Dividing a whole number
   by a unit fraction is the same as
   multiplying the whole number by
   the denominator of the fraction.)
2. 3/4 × 1/2 = 3/8 of a pizza
3. 1/2 + 1/4 = 4/8 + 2/8 = 6/8 of the
   cookies
4. 1/3 ÷ 4 = 1/3 ÷ 12/3 = 1/12 of a
   pie
5. 1/2 ÷ 1/4 = 2/4 ÷ 1/4 = 2 people

Application and Enrichment 12G

1. definitely false (3/7 < 1/2)
   1/2 + 2/5 = 5/10 + 4/10 = 9/10
2. definitely false (2/5 < 1/2)
   1/4 + 1/2 = 2/8 + 4/8 = 6/8 or 3/4
3. definitely false (9/10 > 2/5)
   2/5 − 1/4 = 8/20 − 5/20 = 3/20
4. could be true (5/12 < 2/3)
   2/3 − 1/4 = 8/12 − 3/12 = 5/12

5. 1/3 + 1/5 = 5/15 + 3/15 = 8/15
   makes sense; 8/15 is greater than
   both numbers she started with.
6. 3/4 - 1/2 = 6/8 - 4/8 = 2/8 = 1/4
   makes sense; 1/4 is shorter than
   the distance he started with.
7. 1/2 × 2/3 = 2/6
   Drawing C
8. 1/3 + 1/2 = 2/6 + 3/6 = 5/6
   Drawing D
9. 1/3 ÷ 1/2 = 2/6 ÷ 3/6 =
   2 ÷ 3 = 2/3
   Drawing A
10. 1/2 − 1/3 = 3/6 − 2/6 =
    1/6
    Drawing B

Application and Enrichment 13G

1. 66: 2 × 3 × 11
   84: 2 × 2 × 3 × 7
   GCF = 2 × 3 = 6
   66 ÷ 6 = 11; 6 is a factor of 66.
   84 ÷ 6 = 14; 6 is a factor of 84.
2. 62: 2 × 31
   93: 3 × 31
   GCF = 31 (31 is the only prime
   common to both.)
   62 ÷ 31 = 2; 31 is a factor of 62.
   93 ÷ 31 = 3; 31 is a factor of 93.
3. 40 = 2 × 2 × 2 × 5
   90 = 2 × 3 × 3 × 5
   GCF = 2 × 5 = 10
   40 ÷ 10 = 4; 10 is a factor of 40.
   90 ÷ 10 = 9; 10 is a factor of 90.
4. GCF = 5
   5(3 + 2) = 5(5) = 25
   5(3) + 5(2) = 15 + 10 = 25
5. GCF = 6
   6(3 + 4) = 6(7) = 42
   6(3) + 6(4) = 18 + 24 = 42
6. GCF = 8
   8(4 + 7) = 8(11) = 88
   8(4) + 8(7) = 32 + 56 = 88

7. GCF = 9
$9(5 + 9) = 9(14) = 126$
$9(5) + 9(9) = 45 + 81 = 126$

8. GCF = 3
$3(2 + 13) = 3(15) = 45$
$3(2) + 3(13) = 6 + 39 = 45$

Application and Enrichment 14G

1. $2 \times 3 = 6$ sq in
2. $3 \times 2 = 6$; $3 \times 6 = 18$ sq in
3. $18 \div 6 = 3$
4. five
5. ten
6. $24 \times 10 = 240$ sq in
7. $1/2 \times 24 = 12$; $1/2 \times 240 = 120$ sq in
8. $120 \div 240 = 120/240 = 1/2$
9. one sixth
10. one fourth

Application and Enrichment 15G

1. $[(8) + 1] \times 5 =$
$[9] \times 5 = 45$
2. $[(10) \times 9] + 2 =$
$[90] + 2 = 92$
3. $2 \times [(7) - 5]$
$2 \times [2] = 4$
4. $7 + 4[9 - (6)] =$
$7 + 4[3] =$
$7 + 12 = 19$
5. $[(24 \div 4) + 2] \times 6 =$
$[(6) + 2] \times 6 =$
$[8] \times 6 = 48$
6. $[(15 - 8) \times 2] + 5 =$
$[(7) \times 2] + 5 =$
$[14] + 5 = 19$
7. $[(3 + 3) \times 5] \div 10 =$
$[(6) \times 5] \div 10 =$
$[30] \div 10 = 3$
8. $2/10 = 20/100$
$20/100 + 70/100 = 90/100$
9. $4/10 = 40/100$
$40/100 - 1/100 = 39/100$

10. $2/10 = 20/100$
$10/100 + 20/100 = 30/100$

Application and Enrichment 16G

1. quadrilateral, trapezoid
2. quadrilateral, parallelogram, rhombus
3. quadrilateral, parallelogram, rectangle
4. quadrilateral, parallelogram, rectangle, rhombus, square
5. 4
6. rhombus, square, rectangle; parallelograms

Application and Enrichment 17G

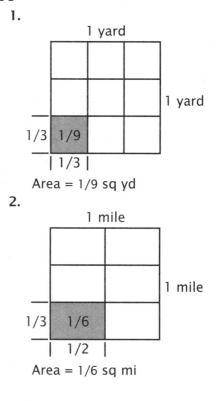

1.

1 yard

1 yard

1/3   1/9

| 1/3 |

Area = 1/9 sq yd

2.

1 mile

1 mile

1/3   1/6

| 1/2 |

Area = 1/6 sq mi

3.

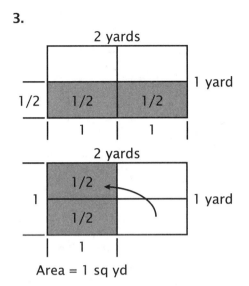

Area = 1 sq yd

Application and Enrichment 18G

1. Main part: $3 \times 4 \times 6 = 72$ cu yd
   Tower: $1 \times 1 \times 3 = 3$ cu yd
   Total: $72 + 3 = 75$ cu yd

2. How the student arrives at the answer to this problem may vary. The floor plan must be divided into rectangular sections to compute the area, and there are multiple correct ways to do so. One example is shown. The final answer should always be the same.

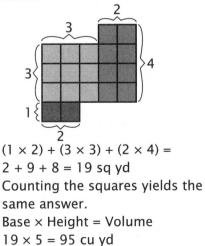

$(1 \times 2) + (3 \times 3) + (2 \times 4) =$
$2 + 9 + 8 = 19$ sq yd
Counting the squares yields the same answer.
Base × Height = Volume
$19 \times 5 = 95$ cu yd

Application and Enrichment 19G

1. 16 leaves
2. $5\frac{1}{2} - 1\frac{1}{2} = 4$ in
   Counting yields the same answer.
3. $1\frac{1}{2} + 4\frac{1}{2} = 6$ in
4. $11/2 \times 3/1 = 33/2 = 16\frac{1}{2}$ in
5. Graph should show one dot over the 1 mark, two dots over the 2 1/2 mark, and three dots over the 4 mark.
6. 4 occurred most; 1 occured least
7. $4 + 4 + 4 = 12$ cups
   $2\frac{1}{2} + 2\frac{1}{2} = 5$ cups
   $12 + 5 + 1 = 18$ cups
8. $18 \div 6 = 3$ cups average

Application and Enrichment 20G

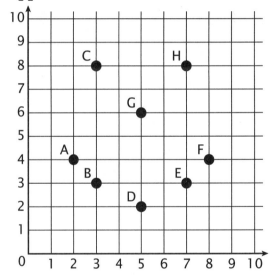

Application and Enrichment 21G

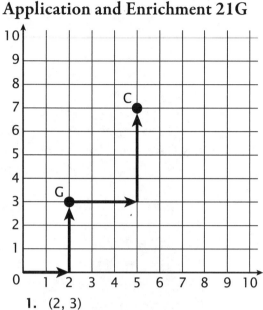

1. (2, 3)
2. (5, 7); 12 blocks
3. The new route will also be twelve blocks as long, as Bill always walks in a direction that brings him closer to home. We are assuming that he stays on the sidewalks and does not cut across any blocks diagonally.
4. 2, 4, 6, 8
5. 0, 4, 8, 12, 16
6. (0, 0), (2, 4), (4, 8), (6, 12), (8, 16)

7.

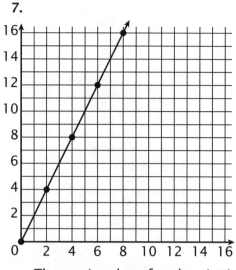

The y-axis value of each point is twice the x-axis value.

Application and Enrichment 22G

1. $\{[(36 \div 6) + 3] - 1\} \times 2 =$
   $\{[6 + 3] - 1\} \times 2 =$
   $\{9 - 1\} \times 2 =$
   $8 \times 2 = 16$

2. $\{[(17 - 9) \times 3] - 2\} \div 11 =$
   $\{[8 \times 3] - 2\} \div 11 =$
   $\{24 - 2\} \div 11 =$
   $22 \div 11 = 2$

3. $2 \times \{[8 + (4 \times 7)] \div 4\} =$
   $2 \times \{[8 + 28] \div 4\} =$
   $2 \times \{36 \div 4\} =$
   $2 \times 9 = 18$

4. $5 + \{[3(6 + 1)] - 4\} =$
   $5 + \{[3(7)] - 4\} =$
   $5 + \{[21] - 4\} =$
   $5 + \{17\} = 22$

5.

2	4	6	8	10	12	14	16
3	6	9	12	15	18	21	24

The student may find and describe patterns other than those given.

6.

1	2	3	4	5	6	7	8
4	8	12	16	20	24	28	32

Each number in the bottom row is four times the number above it.

7.

1	2	4	7	11	16	22	29
0	1	3	6	10	15	21	28

Add a number one greater each time, starting with 1 in the top row and 0 in the bottom.

8.

3	6	9	12	15	18	21	24
24	21	18	15	12	9	6	3

The top row skip counts by three. The bottom row subtracts three each time.

Application and Enrichment 23G

1. 5/6 ÷ 1/6 = 5 ÷ 1 = 5

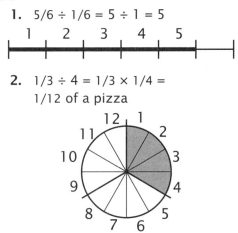

2. 1/3 ÷ 4 = 1/3 × 1/4 = 1/12 of a pizza

3. $3\frac{1}{8} \div 1\frac{1}{4}$ = 25/8 ÷ 5/4 = 25/8 ÷ 10/8 = 25 ÷ 10 = $2\frac{1}{2}$ ft

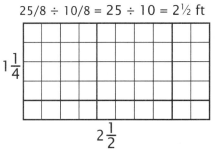

Adding the areas of the parts results in $3\frac{1}{8}$ sq ft.

4. 6 cups ÷ 1/2 cups = 6/1 × 2/1 = 12 times

1	3	5
2	4	6

7	9	11
8	10	12

Application and Enrichment 24G

1. 4 qt
2. multiply
3.

10	40
15	60
8	32
4	16
12	48

4. Tuesday
   40 + 60 + 32 + 16 + 48 = 196 qt
5. $18 × 3 = $54
6. 3 ft
7. divide

8.

2	2/3
3	1
6	2
5	1⅔
1	1/3

9. 1/3 = 8/24
3/8 = 9/24
9/24 > 8/24, so 3/8 > 1/3
3/8 of a yard will be enough.

10. 2/3 = 16/24
1/2 = 12/24
3/8 = 9/24
3/4 = 18/24
Since 18/24 is greater than 16/24, Sarah needs to buy 3/4 of a yard.

Application and Enrichment 25G

1. counting: 15 sq ft
multiplying:
6 × 2½ =
6 × 5/2 = 30/2 = 15 sq ft

2. 1/2 × 2½ = 1/2 × 5/2 =
5/4 = 1¼ pizzas

3. counting: 8¾ sq mi
multiplying:
2½ × 3½ =
5/2 × 7/2 =
35/4 = 8¾ sq mi

4.

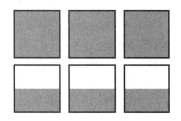

counting:
3 + 1/2 + 1/2 + 1/2 = 4½ cups

multiplying:
1½ × 3 =
3/2 × 3 =
9/2 = 4½ cups

5.

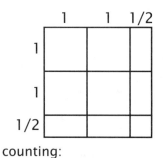

counting:
4 + 4/2 + 1/4 = 6¼ sq yd
multiplying:
2½ × 2½ =
5/2 × 5/2 =
25/4 = 6¼ sq yd

Application and Enrichment 26G

1. drawing and counting: 14 pieces
dividing:
3½ ÷ 1/4 =
7/2 × 4/1 = 28/2 = 14 pieces

2. 2¾ ÷ 1/8 =
11/4 ÷ 1/8 =
11/4 × 8/1 = 88/4 = 22 people

3.

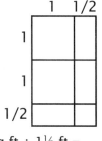

3¾ sq ft ÷ 1½ ft =
15/4 ÷ 3/2 =
15/4 × 2/3 =
30/12 = 2½ ft

4. 4 times

5 ÷ 1¼ =

5 ÷ 5/4 =

5 × 4/5 = 20/5 = 4 times

5. Each gets 2 whole cookies. The remaining 1 1/2 cookies can be evenly divided: 1/2 per person.

7 1/2 ÷ 3 =

15/2 × 1/3 = 15/6 =

2 1/2 cookies

6. 6 2/3 ÷ 1 1/3 =

20/3 ÷ 4/3 =

20 ÷ 4 = 5 sections.

Since there is also a bench at the beginning of the trail, there will be 6 benches. The drawing helps make this clearer.

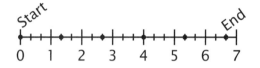

Application and Enrichment 27G

1	2	3	4	5	6	7	8	9	10
2	4	6	8	10	12	14	16	18	20

1. 5 squares
2. 10 triangles
3. skip counting by ones and twos
4. 20 squares, 40 triangles
5.

1	2	3	4	5	6	7	8	9	10
3	6	9	12	15	18	21	24	27	30

6. 15 flies missed
7. 9 flies caught
8.

1	2	3	4	5	6	7	8	9	10
1	2	3	4	5	6	7	8	9	10
2	4	6	8	10	12	14	16	18	20

9. 12 saved cookies
10. 7 + 14 = 21 cookies taken out
11. 20 cookies eaten right away

Application and Enrichment 28G

1.

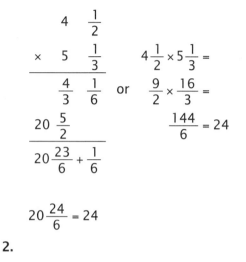

$$20\frac{24}{6} = 24$$

2.

$$
\begin{array}{r}
3 \quad \frac{1}{7} \\
\times \quad 2 \quad \frac{1}{2} \\
\hline
\frac{3}{2} \quad \frac{1}{14} \\
6 \quad \frac{2}{7} \\
\hline
6\frac{28}{14} + \frac{1}{14}
\end{array}
\quad \text{or} \quad
\begin{array}{l}
3\frac{1}{7} \times 2\frac{1}{2} = \\
\frac{22}{7} \times \frac{5}{2} = \\
\frac{110}{14} = \\
7\frac{12}{14} = 7\frac{6}{7}
\end{array}
$$

$$6\frac{26}{14} = 7\frac{12}{14} = 7\frac{6}{7}$$

Application and Enrichment 29G

1. 4.2 > 3.8
2. 0.90 = 0.9
3. 2.31 > 1.31
4. 0.57 < 0.75
5. 0.123 < 0.238
6. 0.8 > 0.12
7. 1.62 > 0.83

Application and Enrichment 30G

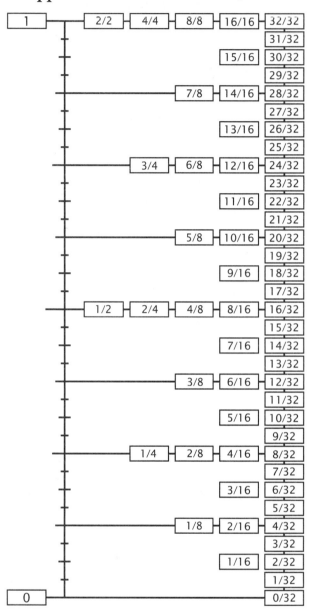

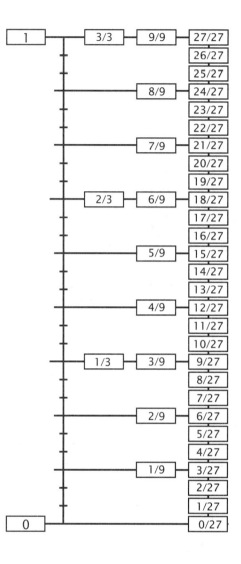

Test Solutions

Lesson Test 1

1. $\frac{1}{2}$ of 12 is 6.
2. $\frac{3}{4}$ of 16 is 12.
3. $15 \div 3 = 5$; $5 \times 1 = 5$
4. $20 \div 5 = 4$; $4 \times 2 = 8$
5. $16 \div 8 = 2$; $2 \times 1 = 2$
6. $8 \div 4 = 2$; $2 \times 3 = 6$
7. $10 \div 2 = 5$; $5 \times 1 = 5$
8. $36 \div 6 = 6$; $6 \times 3 = 18$
9. $49 \div 7 = 7$; $7 \times 2 = 14$
10. $40 \div 5 = 8$; $8 \times 4 = 32$
11. $8 + 5 + 8 + 5 = 26$ in
12. $24 + 12 + 24 + 12 = 72$ ft
13. $55 + 32 + 55 + 32 = 174$ in
14. $18 \div 6 = 3$; $3 \times 1 = 3$ people
15. $\$14.58 + \$6.75 = \$21.33$
16. $\$25.00 - \$21.33 = \$3.67$
17. $14 \div 7 = 2$; $2 \times 2 = 4$ days
18. $12 + 15 + 12 + 15 = 54$ ft

Lesson Test 2

1. $\frac{3}{4}$; three fourths
2. $\frac{1}{5}$; one fifth
3. $\frac{3}{6}$; three sixths
4. $\frac{4}{5}$; 5 sections; 4 shaded
5. $\frac{1}{3}$; 3 sections, 1 shaded
6. one half; 2 sections, 1 shaded
7. $25 \div 5 = 5$; $5 \times 2 = 10$
8. $80 \div 10 = 8$; $8 \times 1 = 8$
9. $42 \div 7 = 6$; $6 \times 3 = 18$
10. $81 + 32 + 81 + 32 = 226$ yd
11. $28 + 28 + 28 + 28 = 112$ ft
12. $\frac{3}{5}$ of 35 = 21 people
13. $10 + 10 + 10 + 10 = 40$ ft
14. $\frac{1}{5}$ of 40 = 8 ft

15. $\$16.45 - \$5.30 = \$11.15$

Lesson Test 3

1. $\frac{1}{6} + \frac{4}{6} = \frac{5}{6}$ or five sixths
2. $\frac{3}{4} - \frac{1}{4} = \frac{2}{4}$ or two fourths
3. $\frac{3}{5} + \frac{2}{5} = \frac{5}{5}$
4. $\frac{4}{6} + \frac{1}{6} = \frac{5}{6}$
5. $\frac{2}{10} + \frac{6}{10} = \frac{8}{10}$
6. $\frac{5}{8} - \frac{3}{8} = \frac{2}{8}$
7. $\frac{8}{9} - \frac{2}{9} = \frac{6}{9}$
8. $\frac{2}{3} - \frac{1}{3} = \frac{1}{3}$
9. $24 \div 8 = 3$; $3 \times 1 = 3$
10. $10 \div 5 = 2$; $2 \times 4 = 8$
11. $20 \div 10 = 2$; $2 \times 4 = 8$
12. $54 + 59 + 98 = 211$ ft
13. $12 + 8 + 12 + 8 = 40$ yd
14. $16 + 16 + 16 + 16 = 64$ in
15. $\frac{1}{6} + \frac{2}{6} = \frac{3}{6}$ of a pie
16. $\frac{3}{3} - \frac{2}{3} = \frac{1}{3}$ of the job
17. $12 \div 4 = 3$; $3 \times 3 = 9$ months
18. $\frac{1}{6} + \frac{2}{6} = \frac{3}{6}$ of the doughnuts

    $12 \div 6 = 2$

    $2 \times 3 = 6$

    $\frac{3}{6}$ of 12 = 6 doughnuts
19. $\frac{6}{6} - \frac{3}{6} = \frac{3}{6}$ of the doughnuts

    $\frac{3}{6}$ of 12 = 6 doughnuts
20. $\$15.50 + \$5.25 + \$10 = \$30.75$

Lesson Test 4

1. $\dfrac{3}{4} = \dfrac{6}{8} = \dfrac{9}{12} = \dfrac{12}{16} = \dfrac{15}{20}$

   three fourths = six eighths = nine twelfths = twelve sixteenths = fifteen twentieths

2. $\dfrac{1}{5} = \dfrac{2}{10} = \dfrac{3}{15} = \dfrac{4}{20}$

3. $\dfrac{5}{6} = \dfrac{10}{12} = \dfrac{15}{18} = \dfrac{20}{24}$

4. $\dfrac{3}{7} = \dfrac{6}{14} = \dfrac{9}{21} = \dfrac{12}{28}$

5. $\dfrac{1}{2} = \dfrac{2}{4} = \dfrac{3}{6} = \dfrac{4}{8}$

6. $\dfrac{4}{5} - \dfrac{1}{5} = \dfrac{3}{5}$

7. $\dfrac{5}{8} + \dfrac{2}{8} = \dfrac{7}{8}$

8. $\dfrac{5}{9} - \dfrac{4}{9} = \dfrac{1}{9}$

9. $48 \div 8 = 6;\ 6 \times 3 = 18$

10. $18 \div 6 = 3;\ 3 \times 1 = 3$

11. $27 \div 3 = 9;\ 9 \times 2 = 18$

12. $22 \times 31 = 682$

13. $54 \times 11 = 594$

14. $322 \times 22 = 7,084$

15. $\dfrac{1}{2} = \dfrac{2}{4}$ of the pages

16. $\dfrac{1}{2}$ of 20 = 10 pages or

    $\dfrac{2}{4}$ of 20 = 10 pages

17. $13 \times 12 = 156$ months

18. $11 + 11 + 11 + 11 = 44$ ft

19. $\dfrac{2}{7} + \dfrac{3}{7} = \dfrac{5}{7}$ read

    $\dfrac{7}{7} - \dfrac{5}{7} = \dfrac{2}{7}$ left

20. $\dfrac{2}{7}$ of 56 = 16

    $56 \div 7 = 8$

    $8 \times 2 = 16$ pages

Lesson Test 5

1. $\dfrac{15}{30} + \dfrac{12}{30} = \dfrac{27}{30}$

2. $\dfrac{6}{12} + \dfrac{4}{12} = \dfrac{10}{12}$

3. $\dfrac{6}{30} + \dfrac{5}{30} = \dfrac{11}{30}$

4. $\dfrac{5}{20} - \dfrac{4}{20} = \dfrac{1}{20}$

5. $\dfrac{25}{30} - \dfrac{12}{30} = \dfrac{13}{30}$

6. $\dfrac{4}{8} - \dfrac{2}{8} = \dfrac{2}{8}$

7. $\dfrac{1}{7} = \dfrac{2}{14} = \dfrac{3}{21} = \dfrac{4}{28}$

8. $\dfrac{2}{3} = \dfrac{4}{6} = \dfrac{6}{9} = \dfrac{8}{12}$

9. $36 \div 9 = 4;\ 4 \times 5 = 20$

10. $90 \div 10 = 9;\ 9 \times 3 = 27$

11. $48 \div 6 = 8;\ 8 \times 4 = 32$

12. 90

13. 10

14. 300

15. 200

16. $\dfrac{3}{18} + \dfrac{6}{18} = \dfrac{9}{18}$ of the soldiers rode

17. $\dfrac{18}{18} - \dfrac{9}{18} = \dfrac{9}{18}$ of the soldiers walked

18. $\dfrac{4}{6} - \dfrac{3}{6} = \dfrac{1}{6}$ of a pound

19. $122 \times 3 = 366$ cups

20. $7 + 24 + 25 = 56$ ft

Lesson Test 6

1. $\dfrac{5}{15} + \dfrac{6}{15} = \dfrac{11}{15}$

2. $\dfrac{18}{24} + \dfrac{4}{24} = \dfrac{22}{24}$

3. $\dfrac{30}{70} + \dfrac{7}{70} = \dfrac{37}{70}$

4. $\dfrac{6}{12} - \dfrac{2}{12} = \dfrac{4}{12}$

5. $\dfrac{15}{24} - \dfrac{8}{24} = \dfrac{7}{24}$

6. $\dfrac{20}{24} - \dfrac{18}{24} = \dfrac{2}{24}$

7. $\dfrac{7}{8} = \dfrac{14}{16} = \dfrac{21}{24} = \dfrac{28}{32}$

8. $\dfrac{1}{3} = \dfrac{2}{6} = \dfrac{3}{9} = \dfrac{4}{12}$

9. $50 \div 10 = 5;\ 5 \times 1 = 5$

10. $16 \div 4 = 4;\ 4 \times 3 = 12$

11. $20 \div 2 = 10;\ 10 \times 1 = 10$

12. $(30) \times (30) = (900)$
$26 \times 33 = 858$

13. $(80) \times (80) = (6,400)$
$75 \times 83 = 6,225$

14. $(60) \times (20) = (1,200)$
$61 \times 17 = 1,037$

15. $16 \times 12 = 192$ books

16. $\frac{1}{2} + \frac{1}{5} = \frac{5}{10} + \frac{2}{10} = \frac{7}{10}$ of his elephants

17. $\frac{4}{9} + \frac{3}{6} = \frac{24}{54} + \frac{27}{54} = \frac{51}{54}$ of the apples

18. $\frac{3}{4} - \frac{1}{8} = \frac{24}{32} - \frac{4}{32} = \frac{20}{32}$ of a pizza

19. 500

20. $\frac{7}{10} \times 60 = 42$ minutes

16. $(70) \times (40) = (2,800)$
$68 \times 43 = 2,924$

17. $\frac{5}{8}$ of $16 = 10$ maple trees

18. $\frac{10}{50} < \frac{15}{50}$ Douglas ate more.

19. $\frac{1}{5}$ of $20 = 4$ chocolates (Christa)

$\frac{3}{10}$ of $20 = 6$ chocolates (Douglas)

$4 < 6$; yes

20. $\frac{8}{10} - \frac{2}{10} = \frac{6}{10}$ inch more

Lesson Test 7

1. $\frac{7}{28} < \frac{12}{28}$

2. $\frac{6}{16} < \frac{8}{16}$

3. $\frac{36}{45} > \frac{10}{45}$

4. $\frac{18}{33} < \frac{22}{33}$

5. $\frac{35}{63} < \frac{54}{63}$

6. $\frac{24}{32} = \frac{24}{32}$

7. $\frac{9}{72} + \frac{32}{72} = \frac{41}{72}$

8. $\frac{10}{15} - \frac{3}{15} = \frac{7}{15}$

9. $\frac{24}{30} + \frac{5}{30} = \frac{29}{30}$

10. $\frac{2}{5} = \frac{4}{10} = \frac{6}{15} = \frac{8}{20}$

11. $19 \div 2 = 9\frac{1}{2}$

12. $51 \div 6 = 8\frac{3}{6}$

13. $39 \div 5 = 7\frac{4}{5}$

14. $(40) \times (20) = (800)$
$39 \times 24 = 936$

15. $(70) \times (20) = (1,400)$
$72 \times 15 = 1,080$

Lesson Test 8

1. $\frac{7}{14} + \frac{6}{14} = \frac{13}{14}$

$\frac{13}{14} + \frac{1}{3} = \frac{39}{42} + \frac{14}{42} = \frac{53}{42} = 1\frac{11}{42}$

2. $\frac{16}{24} + \frac{6}{24} = \frac{22}{24}$

$\frac{22}{24} + \frac{2}{3} = \frac{66}{72} + \frac{48}{72} = \frac{114}{72} = 1\frac{42}{72}$

3. $\frac{24}{32} + \frac{12}{32} = \frac{36}{32}$

$\frac{36}{32} + \frac{1}{2} = \frac{72}{64} + \frac{35}{64} = \frac{104}{64} = 1\frac{40}{64}$

Using the shorter method would yield this solution:

$\frac{6}{8} + \frac{3}{8} + \frac{4}{8} = \frac{13}{8} = 1\frac{5}{8}$

Both solutions have the same value.

4. $\frac{18}{24} + \frac{4}{24} = \frac{22}{24}$

$\frac{22}{24} + \frac{1}{5} = \frac{110}{120} + \frac{24}{120} = \frac{134}{120} = 1\frac{14}{120}$

5. $\frac{4}{16} + \frac{8}{16} = \frac{12}{16}$

$\frac{12}{16} + \frac{2}{5} = \frac{60}{80} + \frac{32}{80} = \frac{92}{80} = 1\frac{12}{80}$

6. $\frac{3}{6} + \frac{4}{6} + \frac{5}{6} = \frac{12}{6} = 2$

7. $\frac{16}{32} = \frac{16}{32}$

8. $\frac{24}{56} > \frac{21}{56}$

9. $\frac{8}{20} < \frac{15}{20}$

10. $\frac{3}{9} = \frac{6}{18} = \frac{9}{27} = \frac{12}{36}$

11. $\dfrac{2}{3} = \dfrac{4}{6} = \dfrac{6}{9} = \dfrac{8}{12}$

12. $298 \div 3 = 99\dfrac{1}{3}$

13. $156 \div 7 = 22\dfrac{2}{7}$

14. $465 \div 4 = 116\dfrac{1}{4}$

15. $\dfrac{2}{8} + \dfrac{4}{8} + \dfrac{1}{8} = \dfrac{7}{8}$ of his plants

16. $\dfrac{8}{8} - \dfrac{7}{8} = \dfrac{1}{8}$ alive. $\dfrac{1}{8}$ of 32 = 4 plants

17. $4 \times 65 = 260$ tomatoes

18. $217 \div 5 = \$43\dfrac{2}{5}$

19. 60

20. 800

Unit Test I

1. $12 \div 3 = 4$; $4 \times 2 = 8$

2. $10 \div 5 = 2$; $2 \times 3 = 6$

3. $24 \div 4 = 6$; $6 \times 3 = 18$

4. $15 \div 3 = 5$; $5 \times 1 = 5$

5. $25 \div 5 = 5$; $5 \times 4 = 20$

6. $14 \div 7 = 2$; $2 \times 5 = 10$

7. $\dfrac{2}{6} = \dfrac{4}{12} = \dfrac{6}{18} = \dfrac{8}{24}$

8. $\dfrac{5}{8} = \dfrac{10}{16} = \dfrac{15}{24} = \dfrac{20}{32}$

9. $\dfrac{1}{3} + \dfrac{1}{3} = \dfrac{2}{3}$

10. $\dfrac{6}{24} + \dfrac{12}{24} = \dfrac{18}{24}$

11. $\dfrac{6}{30} + \dfrac{25}{30} = \dfrac{31}{30} = 1\dfrac{1}{30}$
(final step optional)

12. $\dfrac{10}{15} - \dfrac{3}{15} = \dfrac{7}{15}$

13. $\dfrac{7}{10} - \dfrac{3}{10} = \dfrac{4}{10}$

14. $\dfrac{32}{56} - \dfrac{21}{56} = \dfrac{11}{56}$

15. $\dfrac{21}{35} < \dfrac{25}{35}$

16. $\dfrac{24}{48} = \dfrac{24}{48}$

17. $\dfrac{16}{24} > \dfrac{12}{24}$

18. $\dfrac{30}{80} + \dfrac{8}{80} = \dfrac{38}{80}$

$\dfrac{38}{80} + \dfrac{2}{5} = \dfrac{190}{400} + \dfrac{160}{400} = \dfrac{350}{400}$

19. $\dfrac{6}{8} + \dfrac{7}{8} + \dfrac{4}{8} = \dfrac{17}{8} = 2\dfrac{1}{8}$
(final step optional)

20. $16 + 17 + 28 = 61$ ft

21. $46 + 23 + 46 + 23 = 138$ yd

22. $52 + 52 + 52 + 52 = 208$ in

23. 90

24. 50

25. 600

26. 500

27. $358 \div 5 = 71\dfrac{3}{5}$

28. $541 \div 8 = 67\dfrac{5}{8}$

29. $189 \div 6 = 31\dfrac{3}{6}$

30. $67 \times 18 = 1{,}206$

31. $34 \times 26 = 884$

32. $224 \times 22 = 4{,}928$

33. $\dfrac{2}{5} + \dfrac{1}{3} = \dfrac{6}{15} + \dfrac{5}{15} = \dfrac{11}{15}$ of her money

34. $\dfrac{15}{18} > \dfrac{12}{18}$ so $\dfrac{5}{6} > \dfrac{2}{3}$
Drumore received more snow.

$\dfrac{15}{18} - \dfrac{12}{18} = \dfrac{3}{18}$ ft difference

35. $10 + 11 + 10 + 11 = 42$ ft

$\dfrac{1}{7}$ of 42 = 6 ft of shelves

Lesson Test 9

1. $\dfrac{1}{4}$ of $\dfrac{1}{3} = \dfrac{1}{12}$

2. $\dfrac{3}{4} \times \dfrac{2}{5} = \dfrac{6}{20}$

3. $\dfrac{2}{7} \times \dfrac{1}{4} = \dfrac{2}{28}$

4. $\dfrac{2}{5}$ of $\dfrac{3}{7} = \dfrac{6}{35}$

5. $\dfrac{3}{5} \times \dfrac{1}{6} = \dfrac{3}{30}$

6. $\dfrac{2}{5} \times \dfrac{1}{3} = \dfrac{2}{15}$

7. $\dfrac{12}{44} + \dfrac{11}{44} = \dfrac{23}{44}$

8. $\dfrac{24}{30} - \dfrac{5}{30} = \dfrac{19}{30}$

9. $\dfrac{3}{21} + \dfrac{14}{21} = \dfrac{17}{21}$

   $\dfrac{17}{21} + \dfrac{1}{2} = \dfrac{34}{42} + \dfrac{21}{42} = \dfrac{55}{42}$ or $1\dfrac{13}{42}$

10. $\dfrac{16}{24} > \dfrac{15}{24}$

11. $\dfrac{36}{48} = \dfrac{36}{48}$

12. $\dfrac{45}{54} > \dfrac{42}{54}$

13. $(600) \times (50) = (30,000)$

    $612 \times 54 = 33,048$

14. $(100) \times (40) = (4,000)$

    $124 \times 36 = 4,464$

15. $(1000) \times (10) = (10,000)$

    $957 \times 13 = 12,441$

16. $\dfrac{1}{3} + \dfrac{2}{5} = \dfrac{5}{15} + \dfrac{6}{15} = \dfrac{11}{15}$ of the customers

17. $10 + 12 + 12 = 34$ ft

18. $\dfrac{3}{8} \times \dfrac{1}{2} = \dfrac{3}{16}$ of the people

19. $\dfrac{3}{8}$ of 48;

    $48 \div 8 = 6;\ 6 \times 3 = 18$ burgers

    $\dfrac{1}{2}$ of $\dfrac{3}{8} = \dfrac{3}{16};\ \dfrac{3}{16}$ of 48:

    $48 \div 16 = 3;\ 3 \times 3 = 9$ with mustard

20. $48 \times 15 = \$720$

Lesson Test 10

1. $\dfrac{2}{5} \div \dfrac{1}{5} = \dfrac{2 \div 1}{1} = 2$

2. $\dfrac{15}{24} \div \dfrac{16}{24} = \dfrac{15 \div 16}{1} = \dfrac{15}{16}$

3. $\dfrac{4}{8} \div \dfrac{2}{8} = \dfrac{4 \div 2}{1} = 2$

4. $\dfrac{3}{12} \div \dfrac{4}{12} = \dfrac{3 \div 4}{1} = \dfrac{3}{4}$

5. $\dfrac{27}{45} \div \dfrac{10}{45} = \dfrac{27 \div 10}{1} = \dfrac{27}{10}$ or $2\dfrac{7}{10}$

6. $\dfrac{12}{15} \div \dfrac{10}{15} = \dfrac{12 \div 10}{1} = \dfrac{12}{10}$ or $1\dfrac{2}{10}$

7. $\dfrac{3}{4} \times \dfrac{1}{4} = \dfrac{3}{16}$

8. $\dfrac{2}{3} \times \dfrac{1}{5} = \dfrac{2}{15}$

9. $\dfrac{2}{9} \times \dfrac{1}{2} = \dfrac{2}{18}$

10. $\dfrac{32}{56} + \dfrac{21}{56} = \dfrac{53}{56}$

11. $\dfrac{20}{36} - \dfrac{9}{36} = \dfrac{11}{36}$

12. $\dfrac{4}{10} + \dfrac{3}{10} + \dfrac{5}{10} = \dfrac{12}{10}$ or $1\dfrac{2}{10}$

13. $(500) \div (40) \approx (10)$

14. $(900) \div (30) = (30)$

15. $(600) \div (60) = (10)$

16. $\dfrac{1}{2} \div \dfrac{1}{12} = \dfrac{12}{24} \div \dfrac{2}{24} = \dfrac{12 \div 2}{1} = 6$ pieces

17. $\dfrac{7}{8} \div \dfrac{1}{8} = \dfrac{7 \div 1}{1} = 7$ pieces

18. $\dfrac{3}{10} + \dfrac{1}{10} = \dfrac{4}{10}$

    $\dfrac{4}{10} + \dfrac{3}{5} = \dfrac{20}{50} + \dfrac{30}{50} = \dfrac{50}{50} = 1$ mi

19. $\dfrac{7}{8} - \dfrac{2}{8} = \dfrac{5}{8}$ or $\dfrac{20}{32}$ of a pizza

20. $365 \times 24 = 8,760$ hours

Lesson Test 11

1. no

2. yes

3. no

4. yes

5. $\underline{1}, \underline{2}, \underline{4}, \underline{8}$
   $\underline{1}, \underline{2}, 3, \underline{4}, 6, \underline{8}, 12, 24$
   GCF = 8

6. $\underline{1}, \underline{2}, \underline{5}, \underline{10}$
   $\underline{1}, \underline{2}, 4, \underline{5}, \underline{10}, 20$
   GCF = 10

7. $\underline{1}, \underline{3}, 13, 39$
   $\underline{1}, \underline{3}, 5, 15$
   GCF = 3

8. $\underline{1}, \underline{2}, \underline{7}, \underline{14}$
   $\underline{1}, \underline{2}, 4, \underline{7}, \underline{14}, 28$
   GCF = 14

9. $\dfrac{1}{8} \times \dfrac{2}{7} = \dfrac{2}{56}$

10. $\dfrac{2}{9} \times \dfrac{4}{5} = \dfrac{8}{45}$

11. $\dfrac{3}{6} \times \dfrac{1}{6} = \dfrac{3}{36}$

12. $\dfrac{2}{16} : \dfrac{8}{16} = \dfrac{2 \div 8}{1} = \dfrac{2}{8}$

13. $\dfrac{12}{18} \div \dfrac{3}{18} = \dfrac{12 \div 3}{1} = 4$

14. $\dfrac{50}{60} \div \dfrac{18}{60} = \dfrac{50 \div 18}{1} = \dfrac{50}{18}$ or $2\dfrac{14}{18}$

15. $\dfrac{36}{48} = \dfrac{36}{48}$

16. $\dfrac{30}{80} < \dfrac{32}{80}$

17. $\dfrac{9}{18} < \dfrac{10}{18}$

18. $673 \div 26 = 25\dfrac{23}{26}$

19. $390 \div 82 = 4\dfrac{62}{82}$

20. $768 \div 51 = 15\dfrac{3}{51}$

Lesson Test 12

1. $\dfrac{8 \div 2}{10 \div 2} = \dfrac{4}{5}$

2. $\dfrac{12 \div 4}{20 \div 4} = \dfrac{3}{5}$

3. $\dfrac{27 \div 3}{30 \div 3} = \dfrac{9}{10}$

4. $\dfrac{20 \div 10}{30 \div 10} = \dfrac{2}{3}$

5. $\dfrac{15 \div 5}{25 \div 5} = \dfrac{3}{5}$

6. $\dfrac{36 \div 12}{48 \div 12} = \dfrac{3}{4}$

7. yes

8. no

9. no

10. no

11. $\dfrac{12}{24} + \dfrac{6}{24} = \dfrac{18}{24} \div \dfrac{6}{6} = \dfrac{3}{4}$

12. $\dfrac{12}{54} + \dfrac{9}{54} = \dfrac{21}{54} \div \dfrac{3}{3} = \dfrac{7}{18}$

13. $\dfrac{2}{3} \times \dfrac{5}{6} = \dfrac{10}{18} \div \dfrac{2}{2} = \dfrac{5}{9}$

14. $\dfrac{5}{8} \times \dfrac{1}{10} = \dfrac{5}{80} \div \dfrac{5}{5} = \dfrac{1}{16}$

15. $\dfrac{10}{50} \div \dfrac{40}{50} = \dfrac{10 \div 40}{1} = \dfrac{10}{40} \div \dfrac{10}{10} = \dfrac{1}{4}$

16. $\dfrac{12}{21} \div \dfrac{14}{21} = \dfrac{12 \div 14}{1} = \dfrac{12}{14} \div \dfrac{2}{2} = \dfrac{6}{7}$

17. $573 \times 612 = 350,676$

18. $3,682 \times 694 = 2,555,308$

19. $\dfrac{3}{8} \times \dfrac{1}{2} = \dfrac{3}{16}$ of the wall

20. $\dfrac{6}{8} \div \dfrac{1}{4} = \dfrac{24}{32} \div \dfrac{8}{32} = \dfrac{24 \div 8}{1} = 3$ parts

Lesson Test 13

1. $2 \times 5 \times 5$

2. $2 \times 2 \times 2 \times 2 \times 3$

3. $3 \times 3 \times 7$

4. $3 \times 3 \times 11$

5. $\dfrac{20}{30} = \dfrac{\cancel{2} \times 2 \times \cancel{5}}{\cancel{2} \times 3 \times \cancel{5}} = \dfrac{2}{3}$

6. $\dfrac{63}{81} = \dfrac{\cancel{3} \times \cancel{3} \times 7}{\cancel{3} \times \cancel{3} \times 3 \times 3} = \dfrac{7}{9}$

7. $\dfrac{36}{42} = \dfrac{\cancel{2} \times 2 \times \cancel{3} \times 3}{\cancel{2} \times \cancel{3} \times 7} = \dfrac{6}{7}$

8. $\dfrac{12}{72} + \dfrac{30}{72} = \dfrac{42}{72} \div \dfrac{6}{6} = \dfrac{7}{12}$

9. $\dfrac{30}{42} \div \dfrac{35}{42} = \dfrac{30 \div 35}{1} = \dfrac{30}{35} \div \dfrac{5}{5} = \dfrac{6}{7}$

10. $\dfrac{3}{4} \times \dfrac{1}{3} = \dfrac{3}{12} \div \dfrac{3}{3} = \dfrac{1}{4}$

11. $\dfrac{32}{56} > \dfrac{21}{56}$

12. $\dfrac{55}{110} < \dfrac{60}{110}$

13. $\dfrac{15}{60} > \dfrac{8}{60}$

14. $5311 \div 83 = 63\dfrac{82}{83}$

15. $8856 \div 36 = 246$

16. $6532 \div 512 = 12\dfrac{388}{512}$ or $12\dfrac{97}{128}$

17. $25 \times 17 = 425$ qt

18. $\dfrac{1}{7} + \dfrac{7}{10} = \dfrac{10}{70} + \dfrac{49}{70} = \dfrac{59}{70}$ riding

$\dfrac{70}{70} - \dfrac{59}{70} = \dfrac{11}{70}$ walking

19. $\dfrac{1}{2} \times \dfrac{1}{3} = \dfrac{1}{6}$ with backpacks

20. $\dfrac{12}{42} > \dfrac{7}{42}$ so $\dfrac{2}{7} > \dfrac{1}{6}$

The group that bought dictionaries was larger.

Lesson Test 14

1. $\dfrac{4}{5}$

2. $\dfrac{4}{8} = \dfrac{1}{2}$

3. $\dfrac{10}{16} = \dfrac{5}{8}$

4. $\dfrac{2}{16} = \dfrac{1}{8}$

5. $\dfrac{13}{16}$

6. $\dfrac{5}{16}$

7. $2 \times 2 \times 2 \times 3$

8. $2 \times 2 \times 19$

9. $2 \times 2 \times 2 \times 2 \times 3$

10. $\dfrac{32}{54} = \dfrac{\cancel{2} \times 2 \times 2 \times 2 \times 2}{\cancel{2} \times 3 \times 3 \times 3} = \dfrac{16}{27}$

11. $32 \times 15 = 480$ sq in

12. $10 \times 4 = 40$ sq ft

13. $49 \times 24 = 1{,}176$ sq yd

14. no

15. $16 : \underline{2}, \underline{4}, 8, 16$
    $44 : \underline{2}, \underline{4}, 11, 22, 44$
    GCF = 4

16. $\dfrac{2}{5} \times \dfrac{3}{4} = \dfrac{6}{20} \div \dfrac{2}{2} = \dfrac{3}{10}$ of the days

17. $650 \div 10 = 65$ bags

18. perimeter

Lesson Test 15

1. $\dfrac{9}{9} + \dfrac{2}{9} = \dfrac{11}{9}$

2. $\dfrac{5}{5} + \dfrac{5}{5} + \dfrac{5}{5} + \dfrac{3}{5} = \dfrac{18}{5}$

3. $\dfrac{3}{3} + \dfrac{1}{3} = \dfrac{4}{3}$

4. $\dfrac{4}{4} + \dfrac{4}{4} + \dfrac{1}{4} = \dfrac{9}{4}$

5. $\dfrac{4}{4} + \dfrac{3}{4} = 1\dfrac{3}{4}$

6. $\dfrac{2}{2} + \dfrac{2}{2} + \dfrac{2}{2} + \dfrac{2}{2} + \dfrac{1}{2} = 4\dfrac{1}{2}$

7. $\dfrac{5}{5} + \dfrac{5}{5} + \dfrac{3}{5} = 2\dfrac{3}{5}$

8. $\dfrac{6}{6} + \dfrac{6}{6} + \dfrac{6}{6} + \dfrac{5}{6} = 3\dfrac{5}{6}$

9. $\dfrac{14}{16} = \dfrac{7}{8}$

10. $\dfrac{8}{16} = \dfrac{1}{2}$

11. $\dfrac{3}{9} \div \dfrac{3}{3} = \dfrac{1}{3}$

12. $\dfrac{20}{28} \div \dfrac{4}{4} = \dfrac{5}{7}$

13. $\dfrac{18}{32} \div \dfrac{2}{2} = \dfrac{9}{16}$

14. $16^2 = 256$ sq in

15. $3^2 = 9$ sq mi

16. $8^2 = 64$
    (See Quick Tip on 15E in student book.)

17. $6^2 = 36$

18. yes

19. $15 : \underline{3}, \underline{5}, \underline{15}$
    $30 : 2, \underline{3}, \underline{5}, 6, 10, \underline{15}, 30$
    GCF = 15

20. $3 \times 3 \times 5$

Lesson Test 16

1. $3\dfrac{3}{4}$

2. $2\dfrac{8}{10} = 2\dfrac{4}{5}$

3. $\dfrac{8}{8} + \dfrac{1}{8} = \dfrac{9}{8}$

4. $\dfrac{6}{6} + \dfrac{6}{6} + \dfrac{5}{6} = \dfrac{17}{6}$

5. $\dfrac{4}{4} + \dfrac{4}{4} + \dfrac{1}{4} = 2\dfrac{1}{4}$

6. $\dfrac{7}{7} + \dfrac{7}{7} + \dfrac{7}{7} + \dfrac{3}{7} = 3\dfrac{3}{7}$

7. $\dfrac{27}{63} + \dfrac{14}{63} = \dfrac{41}{63}$

8. $\dfrac{6}{48} + \dfrac{8}{48} = \dfrac{14}{48} \div \dfrac{2}{2} = \dfrac{7}{24}$

9. $\dfrac{8}{24} - \dfrac{6}{24} = \dfrac{2}{24} \div \dfrac{2}{2} = \dfrac{1}{12}$

10. $\dfrac{5}{6} \times \dfrac{11}{12} = \dfrac{55}{72}$

11. $\dfrac{10}{20} \div \dfrac{6}{20} = \dfrac{10 \div 6}{20_1} = \dfrac{10}{6} = \dfrac{5}{3} = 1\dfrac{2}{3}$

12. $\dfrac{2}{3} \times \dfrac{7}{20} = \dfrac{14}{60} \div \dfrac{2}{2} = \dfrac{7}{30}$

13. $10 \times 4 = 40$ sq ft

14. $10 + 4 + 10 + 4 = 28$ ft

15. $23^2 = 529$ sq in

16. $23 + 23 + 23 + 23 = 92$ in

17. $\left(\dfrac{1}{2}\right)(12)(5) = 30$ sq ft

18. $5 + 12 + 13 = 30$ ft

19. $16 \times 16 = 256$

20. $36 : \underline{2}, \underline{3}, 4, \underline{6}, 9, 12, 18, 36$
$42 : \underline{2}, \underline{3}, \underline{6}, 7, 14, 21, 42$
GCF = 6

Unit Test II

1. $\dfrac{9}{18} \div \dfrac{10}{18} = \dfrac{9 \div 10}{1} = \dfrac{9}{10}$

2. $\dfrac{18}{42} \div \dfrac{35}{42} = \dfrac{18 \div 35}{1} = \dfrac{18}{35}$

3. $\dfrac{56}{98} \div \dfrac{35}{98} = \dfrac{56 \div 35}{1} = \dfrac{56}{35} = \dfrac{8}{5} = 1\dfrac{3}{5}$

4. $\dfrac{21}{27} \div \dfrac{18}{27} = \dfrac{21 \div 18}{1} = \dfrac{21}{18} = \dfrac{7}{6} = 1\dfrac{1}{6}$

5. $\dfrac{24}{32} \div \dfrac{4}{32} = \dfrac{24 \div 4}{1} = 6$

6. $\dfrac{5}{20} \div \dfrac{12}{20} = \dfrac{5 \div 12}{1} = \dfrac{5}{12}$

7. $\dfrac{1}{4} \times \dfrac{1}{3} = \dfrac{1}{12}$

8. $\dfrac{4}{5} \times \dfrac{7}{10} = \dfrac{28}{50} = \dfrac{14}{25}$

9. $\dfrac{2}{5} \times \dfrac{1}{2} = \dfrac{2}{10} = \dfrac{1}{5}$

10. $\dfrac{3}{4} \times \dfrac{1}{4} = \dfrac{3}{16}$

11. $\dfrac{3}{7} \times \dfrac{2}{9} = \dfrac{6}{63} = \dfrac{2}{21}$

12. $\dfrac{1}{8} \times \dfrac{1}{6} = \dfrac{1}{48}$

13. yes

14. no

15. yes

16. no

17. $28 : \underline{2}, 4, 7, 14, 28$
$54 : \underline{2}, 3, 6, 9, 18, 27, 54$
GCF = 2

18. $2 \times 2 \times 3 \times 5$

19. $\dfrac{4}{12} \div \dfrac{4}{4} = \dfrac{1}{3}$

20. $\dfrac{15}{20} \div \dfrac{5}{5} = \dfrac{3}{4}$

21. $\dfrac{18}{42} \div \dfrac{6}{6} = \dfrac{3}{7}$

22. $\dfrac{4}{4} + \dfrac{3}{4} = \dfrac{7}{4}$

23. $\dfrac{9}{9} + \dfrac{9}{9} + \dfrac{7}{9} = \dfrac{25}{9}$

24. $\dfrac{5}{5} + \dfrac{5}{5} + \dfrac{3}{5} = 2\dfrac{3}{5}$

25. $\dfrac{3}{3} + \dfrac{3}{3} + \dfrac{3}{3} + \dfrac{1}{3} = 3\dfrac{1}{3}$

26. $1\dfrac{10}{16} = 1\dfrac{5}{8}$

27. $4\dfrac{1}{10}$

28. $3\dfrac{2}{8} = 3\dfrac{1}{4}$

29. $7 \times 3 = 21$ sq ft

30. $12^2 = 144$ sq in

31. $\left(\dfrac{1}{2}\right)(8)(6) = 24$ sq ft

32. $9 \times 9 = 81$

33. $\dfrac{3}{4} \div \dfrac{1}{16} = \dfrac{48}{64} \div \dfrac{4}{64} = \dfrac{48 \div 4}{1} = 12$ people

34. $\dfrac{3}{4} \times \dfrac{1}{2} = \dfrac{3}{8}$ of a cake

Lesson Test 17

1. $2\dfrac{1}{10} + 6\dfrac{6}{10} = 8\dfrac{7}{10}$

2. $4\dfrac{3}{5} - 3\dfrac{1}{5} = 1\dfrac{2}{5}$

3. $5\dfrac{2}{7} + 1\dfrac{4}{7} = 6\dfrac{6}{7}$

4. $6\dfrac{1}{2} = \dfrac{13}{2}$

5. $5\dfrac{2}{3} = \dfrac{17}{3}$

6. $1\dfrac{9}{10} = \dfrac{19}{10}$

7. $\dfrac{21}{4} = 5\dfrac{1}{4}$

8. $\dfrac{11}{5} = 2\dfrac{1}{5}$

9. $\dfrac{15}{8} = 1\dfrac{7}{8}$

10. $\dfrac{5}{6} \times \dfrac{1}{3} = \dfrac{5}{18}$

11. $\dfrac{1}{10} \div \dfrac{3}{10} = \dfrac{1 \div 3}{1} = \dfrac{1}{3}$

12. $\dfrac{8}{9} \times \dfrac{5}{6} = \dfrac{40}{54} = \dfrac{20}{27}$

13. $5 \times 5 \times 5 = 125$ cu in

14. $8 \times 8 \times 8 = 512$ cu ft

15. $13 \times 13 \times 13 = 2{,}197$ cu ft

16. $17 \times 17 \times 17 = 4{,}913$ cu ft

17. 5×19

18. $4\dfrac{3}{10} + 2\dfrac{6}{10} = 6\dfrac{9}{10}$ lb

19. $\dfrac{3}{4} \times \dfrac{3}{4} = \dfrac{9}{16}$ sq mi

20. $\dfrac{5}{8}$ of 120

$120 \div 8 = 15$; $15 \times 5 = 75$ parents

Lesson Test 18

1. $6\dfrac{4}{7} + 7\dfrac{5}{7} = 13\dfrac{9}{7} = 14\dfrac{2}{7}$

2. $13\dfrac{4}{9} + 5\dfrac{8}{9} = 18\dfrac{12}{9} = 19\dfrac{3}{9} = 19\dfrac{1}{3}$

3. $21\dfrac{11}{12} + 3\dfrac{1}{12} = 24\dfrac{12}{12} = 25$

4. $4\dfrac{3}{5} + 8\dfrac{4}{5} = 12\dfrac{7}{5} = 13\dfrac{2}{5}$

5. $9\dfrac{2}{3} + 8\dfrac{2}{3} = 17\dfrac{4}{3} = 18\dfrac{1}{3}$

6. $6\dfrac{7}{8} + 7\dfrac{3}{8} = 13\dfrac{10}{8} = 14\dfrac{2}{8} = 14\dfrac{1}{4}$

7. $\dfrac{3}{4} - \dfrac{1}{5} = \dfrac{15}{20} - \dfrac{4}{20} = \dfrac{11}{20}$

8. $\dfrac{1}{10} + \dfrac{1}{5} = \dfrac{5}{50} + \dfrac{10}{50} = \dfrac{15}{50} = \dfrac{3}{10}$

9. $\dfrac{7}{9} - \dfrac{1}{2} = \dfrac{14}{18} - \dfrac{9}{18} = \dfrac{5}{18}$

10. $\dfrac{3}{10} \div \dfrac{1}{10} = \dfrac{3 \div 1}{1} = 3$

11. $\dfrac{5}{6} \times \dfrac{1}{3} = \dfrac{5}{18}$

12. $\dfrac{3}{4} \div \dfrac{1}{3} = \dfrac{9}{12} \div \dfrac{4}{12} = \dfrac{9 \div 4}{1} = \dfrac{9}{4} = 2\dfrac{1}{4}$

13. $64 \div 8 = 8$; $8 \times 3 = 24$

14. $42 \div 6 = 7$; $7 \times 1 = 7$

15. $80 \div 10 = 8$; $8 \times 7 = 56$

16. $8 \times 5 \times 2 = 80$ cu ft

17. $80 \times 63 = 5{,}040$ lb

18. $42 : \underline{2}, 3, 6, \underline{7}, \underline{14}, 21, 42$
    $56 : \underline{2}, 4, \underline{7}, 8, \underline{14}, 28, 56$
    $GCF = 14$

19. $2 \times 3 \times 3$

20. $2\dfrac{4}{5} + 1\dfrac{3}{5} = 3\dfrac{7}{5} = 4\dfrac{2}{5}$ mi

Lesson Test 19

1. $2\dfrac{1}{8} - 1\dfrac{5}{8} = 1\dfrac{9}{8} - 1\dfrac{5}{8} = \dfrac{4}{8} = \dfrac{1}{2}$

2. $5\dfrac{2}{7} - \dfrac{3}{7} = 4\dfrac{9}{7} - \dfrac{3}{7} = 4\dfrac{6}{7}$

3. $3 - 2\dfrac{1}{4} = 2\dfrac{4}{4} - 2\dfrac{1}{4} = \dfrac{3}{4}$

4. $2\dfrac{5}{8} + 2\dfrac{4}{8} = 4\dfrac{9}{8} = 5\dfrac{1}{8}$

5. $7\dfrac{3}{5} + 4\dfrac{4}{5} = 11\dfrac{7}{5} = 12\dfrac{2}{5}$

6. $8\dfrac{3}{6} + 9\dfrac{3}{6} = 17\dfrac{6}{6} = 18$

7. $\dfrac{4}{5} \div \dfrac{2}{3} = \dfrac{12}{15} \div \dfrac{10}{15} = \dfrac{12 \div 10}{1} =$
    $\dfrac{12}{10} = 1\dfrac{2}{10} = 1\dfrac{1}{5}$

8. $\dfrac{2}{11} \times \dfrac{1}{5} = \dfrac{2}{55}$

9. $\dfrac{1}{2} + \dfrac{2}{3} + \dfrac{5}{6} = \dfrac{3}{6} + \dfrac{4}{6} + \dfrac{5}{6} = \dfrac{12}{6} = 2$

10. $\dfrac{5}{20} < \dfrac{8}{20}$

11. $\dfrac{24}{56} > \dfrac{7}{56}$

12. $\dfrac{50}{120} > \dfrac{48}{120}$

13. $99 \div 3 = 33$ yd

14. $26 \times 3 = 78$ ft

15. $9 \times 3 = 27$ ft

16. $\dfrac{2}{3} \times \dfrac{1}{3} = \dfrac{2}{9}$ sq mi

17. $\dfrac{2}{3} + \dfrac{2}{3} + \dfrac{1}{3} + \dfrac{1}{3} = \dfrac{6}{3} = 2$ mi

18. $9\frac{5}{8} + 6\frac{4}{8} = 15\frac{9}{8} = 16\frac{1}{8}$ pies

19. $14\frac{9}{8} - 11\frac{3}{8} = 3\frac{6}{8} = 3\frac{3}{4}$ yd

20. $6 \div 3 = 2$ yd

17. no

18. $9 \div 3 = 3;\ 3 \times 3 \times 3 = 27$ cu yd

19. $2\frac{1}{4} + 2\frac{1}{4} + 2\frac{1}{4} + 2\frac{1}{4} = 8\frac{4}{4} = 9$ ft

20. $15 \times 15 = 225$

Lesson Test 20

1.
$$8\frac{5}{8} + \frac{1}{8} = 8\frac{6}{8}$$
$$-3\frac{7}{8} + \frac{1}{8} = 4$$
$$= 4\frac{6}{8} = 4\frac{3}{4}$$

2.
$$14\ \ + \frac{2}{3} = 14\frac{2}{3}$$
$$-10\frac{1}{3} + \frac{2}{3} = 11$$
$$= 3\frac{2}{3}$$

3. $2\frac{3}{4} + 3\frac{3}{4} = 5\frac{6}{4} = 6\frac{2}{4} = 6\frac{1}{2}$

4. $7\frac{1}{3} - 3\frac{2}{3} = 6\frac{4}{3} - 3\frac{2}{3} = 3\frac{2}{3}$

5. $4\frac{3}{5} + 1\frac{4}{5} = 5\frac{7}{5} = 6\frac{2}{5}$

6. $\frac{2}{3} + \frac{3}{4} = \frac{8}{12} + \frac{9}{12} = \frac{17}{12} = 1\frac{5}{12}$

7. $\frac{3}{4} + \frac{5}{6} = \frac{18}{24} + \frac{20}{24} = \frac{38}{24} = \frac{19}{12} = 1\frac{7}{12}$

8. $\frac{1}{2} + \frac{4}{7} = \frac{7}{14} + \frac{8}{14} = \frac{15}{14} = 1\frac{1}{14}$

9. $\frac{7}{8} - \frac{1}{3} = \frac{21}{24} - \frac{8}{24} = \frac{13}{24}$

10. $\frac{5}{12} - \frac{2}{10} = \frac{50}{120} - \frac{24}{120}$
$$= \frac{26}{120} = \frac{13}{60}$$

11. $\frac{1}{6} - \frac{1}{7} = \frac{7}{42} - \frac{6}{42} = \frac{1}{42}$

12. $32 \times 2 = 64$ pt

13. $34 \div 2 = 17$ qt

14. $7 \times 3 = 21$ ft

15. $9\frac{3}{4} + 6\frac{3}{4} = 15\frac{6}{4} = 16\frac{2}{4} = 16\frac{1}{2}$ tons

16. $4\frac{2}{5} - 1\frac{4}{5} = 3\frac{7}{5}$
$$3\frac{7}{5} - 1\frac{4}{5} = 2\frac{3}{5}$$ gal

Lesson Test 21

1. $9\frac{4}{12} + 6\frac{3}{12} = 15\frac{7}{12}$

2. $4\frac{2}{3} + 1\frac{3}{5} = 4\frac{10}{15} + 1\frac{9}{15} = 5\frac{19}{15} = 6\frac{4}{15}$

3. $9\frac{3}{5} + 2\frac{7}{10} = 9\frac{30}{50} + 2\frac{35}{50} = 11\frac{65}{50}$
$$= 12\frac{15}{50} = 12\frac{3}{10}$$

4. $12\frac{5}{11} + 4\frac{5}{8} = 12\frac{40}{88} + 4\frac{55}{88}$
$$= 16\frac{95}{88} = 17\frac{7}{88}$$

5. $5\frac{1}{5} - 2\frac{3}{5} = 4\frac{6}{5} - 2\frac{3}{5} = 2\frac{3}{5}$

6. $15\frac{7}{9} - 6\frac{2}{9} = 9\frac{5}{9}$

7. $7 - 3\frac{1}{5} = 6\frac{5}{5} - 3\frac{1}{5} = 3\frac{4}{5}$

8. $\frac{1}{2} \div \frac{1}{4} = \frac{4}{8} \div \frac{2}{8} = \frac{4 \div 2}{1} = 2$

9. $\frac{8}{12} \div \frac{1}{3} = \frac{24}{36} \div \frac{12}{36} = \frac{24 \div 12}{1} = 2$

10. $\frac{3}{5} \div \frac{1}{10} = \frac{30}{50} \div \frac{5}{50} = \frac{30 \div 5}{1} = 6$

11. $\frac{4}{6} \times \frac{3}{5} = \frac{12}{30} = \frac{2}{5}$

12. $\frac{1}{2} \times \frac{1}{3} = \frac{1}{6}$

13. $\frac{3}{4} \times \frac{1}{6} = \frac{3}{24} = \frac{1}{8}$

14. $81 \div 3 = 27$ yd

15. $54 \div 2 = 27$ qt

16. $18 \times 2 = 36$ pt

17. $92 \div 3 = 30\frac{2}{3}$ yd

18. $20\frac{4}{8} + 2\frac{6}{8} = 22\frac{10}{8} = 23\frac{2}{8} = 23\frac{1}{4}$ mi

19. A $= \frac{2}{3} \times \frac{2}{3} = \frac{4}{9}$ sq ft

P $= \frac{2}{3} + \frac{2}{3} + \frac{2}{3} + \frac{2}{3} = \frac{8}{3} = 2\frac{2}{3}$ ft

20. $3 \times 4 = 12$
$12 \times 2 = 24$
$24 + 1 = 25$
$25 \div 5 = 5$

18. $16 \times 3 = 48$ ft

19. $14 : \underline{2}, \underline{7}, \underline{14}$
$56 : \underline{2}, 4, \underline{7}, 8, \underline{14}, 28, 56$
GCF = 14

20. $9 + 1 = 10$
$10 - 2 = 8$
$8 \times 7 = 56; \ 56 + 4 = 60$

Lesson Test 22

1. $4\frac{1}{4} - 2\frac{3}{4} = 3\frac{5}{4} - 2\frac{3}{4} = 1\frac{2}{4} = 1\frac{1}{2}$

2. $4\frac{1}{2} - 1\frac{1}{3} = 4\frac{3}{6} - 1\frac{2}{6} = 3\frac{1}{6}$

3. $8\frac{5}{9} - 3\frac{2}{9} = 5\frac{3}{9} = 5\frac{1}{3}$

4. $16\frac{3}{10} - 5\frac{2}{5} = 16\frac{15}{50} - 5\frac{20}{50}$
$= 15\frac{65}{50} - 5\frac{20}{50}$
$= 10\frac{45}{50} = 10\frac{9}{10}$

5. $1\frac{1}{2} + 3\frac{2}{3} = 1\frac{3}{6} + 3\frac{4}{6} = 4\frac{7}{6} = 5\frac{1}{6}$

6. $4\frac{2}{3} + 3\frac{4}{5} = 4\frac{10}{15} + 3\frac{12}{15} = 7\frac{22}{15} = 8\frac{7}{15}$

7. $6\frac{3}{4} + 9\frac{7}{8} = 6\frac{24}{32} + 9\frac{28}{32} = 15\frac{52}{32}$
$= 15\frac{13}{8} = 16\frac{5}{8}$

8. $\frac{5}{16} \div \frac{3}{4} = \frac{20}{64} \div \frac{48}{64} = \frac{20 \div 48}{1}$
$= \frac{20}{48} = \frac{5}{12}$

9. $\frac{1}{2} \times \frac{2}{3} = \frac{2}{6} = \frac{1}{3}$

10. $\frac{7}{8} \div \frac{5}{12} = \frac{84}{96} \div \frac{40}{96} = \frac{84 \div 40}{1}$
$= \frac{84}{40} = \frac{21}{10} = 2\frac{1}{10}$

11. $9\frac{5}{8} = \frac{77}{8}$

12. $25\frac{2}{3} = \frac{77}{3}$

13. $10\frac{3}{4} = \frac{43}{4}$

14. $14 \div 4 = 3\frac{1}{2}$ gal

15. $8 \times 4 = 32$ qt

16. $20 \div 2 = 10$ qt

17. $672 \div 16 = 42$ yd

Lesson Test 23

1. $\frac{3}{4} \div \frac{1}{2} = \frac{3}{4} \times \frac{2}{1} = \frac{6}{4} = 1\frac{2}{4} = 1\frac{1}{2}$

2. $\frac{6}{8} \div \frac{4}{8} = \frac{6 \div 4}{1} = \frac{6}{4} = 1\frac{2}{4} = 1\frac{1}{2}$

3. $\frac{4}{5} \div \frac{2}{3} = \frac{4}{5} \times \frac{3}{2} = \frac{12}{10} = 1\frac{2}{10} = 1\frac{1}{5}$

4. $\frac{12}{15} \div \frac{10}{15} = \frac{12 \div 10}{1} = \frac{12}{10} = 1\frac{2}{10} = 1\frac{1}{5}$

5. $\frac{5}{3} \div \frac{5}{11} = \frac{5}{3} \times \frac{11}{5} = \frac{55}{15} = 3\frac{10}{15} = 3\frac{2}{3}$

6. $\frac{15}{8} \div \frac{1}{8} = \frac{15}{8} \times \frac{8}{1} = \frac{120}{8} = 15$

7. $\frac{11}{4} \div \frac{1}{2} = \frac{11}{4} \times \frac{2}{1} = \frac{22}{4} = 5\frac{2}{4} = 5\frac{1}{2}$

8. $\frac{31}{5} \div \frac{27}{8} = \frac{31}{5} \times \frac{8}{27} = \frac{248}{135} = 1\frac{113}{135}$

9. $20\frac{4}{20} - 10\frac{15}{20} = 19\frac{24}{20} - 10\frac{15}{20} = 9\frac{9}{20}$

10. $9\frac{3}{24} - 4\frac{8}{24} = 8\frac{27}{24} - 4\frac{8}{24} = 4\frac{19}{24}$

11. $5\frac{12}{28} + 5\frac{7}{28} = 10\frac{19}{28}$

12. $17 \div 16 = \frac{17}{16} = 1\frac{1}{16}$ lb

13. $28 \div 2 = 14$ qt

14. $10 \times 3 = 30$ ft

15. $5 \times 16 = 80$ oz

16. $9 \times 2 = 18$ pt

17. $19 \div 4 = \frac{19}{4} = 4\frac{3}{4}$ gal

18. $\frac{25}{8} \div \frac{5}{8} = \frac{25 \div 5}{1} = 5$ people

19. $4 \times 2 \times 1 = 8$ cu ft

20. $6 \times 7 = 42$
$42 + 3 = 45$
$45 \div 9 = 5$
$5 + 2 = 7$

Unit Test III

1. $3\frac{3}{8} + 4\frac{1}{8} = 7\frac{4}{8} = 7\frac{1}{2}$

2. $7\frac{3}{5} + 4\frac{4}{5} = 11\frac{7}{5} = 12\frac{2}{5}$

3. $15\frac{4}{14} + 11\frac{7}{14} = 26\frac{11}{14}$

4. $8\frac{15}{18} + 5\frac{12}{18} = 13\frac{27}{18} = 14\frac{9}{18} = 14\frac{1}{2}$

5. $9\frac{7}{8} - 3\frac{5}{8} = 6\frac{2}{8} = 6\frac{1}{4}$

6. $2\frac{5}{5} - 1\frac{2}{5} = 1\frac{3}{5}$

7. $35\frac{3}{12} - 21\frac{8}{12} = 34\frac{15}{12} - 21\frac{8}{12} = 13\frac{7}{12}$

8. $7\frac{7}{42} - 6\frac{18}{42} = 6\frac{49}{42} - 6\frac{18}{42} = \frac{31}{42}$

9. $48 \div 16 = 3$ lb

10. $23 \div 2 = 11\frac{1}{2}$ qt

11. $9 \times 3 = 27$ ft

12. $4 \times 16 = 64$ oz

13. $16 \times 2 = 32$ pt

14. $20 \div 4 = 5$ gal

15. $\frac{3}{6} \div \frac{1}{2} = \frac{3}{6} \times \frac{2}{1} = \frac{6}{6} = 1$

16. $\frac{6}{12} \div \frac{6}{12} = \frac{6 \div 6}{1} = 1$

17. $\frac{5}{8} \div \frac{2}{3} = \frac{5}{8} \times \frac{3}{2} = \frac{15}{16}$

18. $\frac{15}{24} \div \frac{16}{24} = \frac{15 \div 16}{1} = \frac{15}{16}$

19. $\frac{19}{3} \div \frac{19}{8} = \frac{19}{3} \times \frac{8}{19} = \frac{152}{57} = \frac{8}{3} = 2\frac{2}{3}$

20. $\frac{13}{5} \div \frac{19}{10} = \frac{13}{5} \times \frac{10}{19} = \frac{130}{95} = \frac{26}{19} = 1\frac{7}{19}$

21. $\frac{35}{6} \div \frac{5}{6} = \frac{35}{6} \times \frac{6}{5} = \frac{210}{30} = 7$

22. $\frac{7}{2} \div \frac{10}{7} = \frac{7}{2} \times \frac{7}{10} = \frac{49}{20} = 2\frac{9}{20}$

23. $\frac{1}{8}$

24. $15 \times 10 \times 6 = 900$ cu ft

25. $\frac{9}{2} \div \frac{8}{1} = \frac{9}{2} \times \frac{1}{8} = \frac{9}{16}$ of a pizza

26. $2\frac{6}{18} + 1\frac{15}{18} = 3\frac{21}{18} = 4\frac{3}{18} = 4\frac{1}{6}$ mi

27. $5\frac{3}{6} - 2\frac{4}{6} = 4\frac{9}{6} - 2\frac{4}{6} = 2\frac{5}{6}$ rows

28. $8 \times 6 = 48$; $48 - 4 = 44$; $44 \div 2 = 22$; $22 + 3 = 25$

Lesson Test 24

1. $\frac{1}{10} \cdot 10J = 200 \cdot \frac{1}{10}$
$J = 20$

2. $10(20) = 200$
$200 = 200$

3. $\frac{1}{6} \cdot 6U = 24 \cdot \frac{1}{6}$
$U = 4$

4. $6(4) = 24$
$24 = 24$

5. $\frac{1}{5} \cdot 5K = 95 \cdot \frac{1}{5}$
$K = 19$

6. $5(19) = 95$
$95 = 95$

7. $\frac{1}{3} \div \frac{2}{5} = \frac{1}{3} \times \frac{5}{2} = \frac{5}{6}$

8. $\frac{5}{15} \div \frac{6}{15} = \frac{5 \div 6}{1} = \frac{5}{6}$

9. $\frac{5}{4} \div \frac{7}{2} = \frac{5}{4} \times \frac{2}{7} = \frac{10}{28} = \frac{5}{14}$

10. $\frac{25}{6} \div \frac{24}{5} = \frac{25}{6} \times \frac{5}{24} = \frac{125}{144}$

11. $2\frac{4}{3} - 1\frac{2}{3} = 1\frac{2}{3}$

12. $8\frac{8}{12} + 5\frac{9}{12} = 13\frac{17}{12} = 14\frac{5}{12}$

13. $4\frac{2}{12} - 1\frac{6}{12} = 3\frac{14}{12} - 1\frac{6}{12} = 2\frac{8}{12} = 2\frac{2}{3}$

14. $\frac{2}{3} \times \frac{3}{5} = \frac{6}{15} = \frac{2}{5}$

15. $\frac{2}{7} \times \frac{3}{4} = \frac{6}{28} = \frac{3}{14}$

16. $\frac{5}{8} \times \frac{1}{3} = \frac{5}{24}$

17. $6 \times 12 = 72$ in

18. $3J = 120$; $J = 40$ years

19. $\frac{1}{2} \times 12 = 6$ in

20. $A = \frac{2}{3} \times \frac{2}{3} = \frac{4}{9}$ sq mi
$P = \frac{2}{3} + \frac{2}{3} + \frac{2}{3} + \frac{2}{3} = \frac{8}{3} = 2\frac{2}{3}$ mi

Lesson Test 25

1. $\dfrac{\cancel{2}}{3} \times \dfrac{3}{\cancel{4}} \times \dfrac{\cancel{2}}{7} = \dfrac{1}{7}$

2. $\dfrac{\cancel{12}}{\cancel{7}} \times \dfrac{1}{5} \times \dfrac{\cancel{7}}{\cancel{4}} = \dfrac{3}{5}$

3. $\dfrac{1}{\cancel{8}} \times \dfrac{\cancel{8}}{7} \times \dfrac{4}{7} = \dfrac{4}{49}$

4. $\dfrac{\cancel{6}}{\cancel{5}} \times \dfrac{\cancel{5}}{\cancel{6}} \times \dfrac{3}{2} = \dfrac{3}{2} = 1\dfrac{1}{2}$

5. $\dfrac{1}{7} \cdot 7L = 77 \cdot \dfrac{1}{7}$
   $L = 11$

6. $7(11) = 77$
   $77 = 77$

7. $\dfrac{1}{13} \cdot 13Y = 195 \cdot \dfrac{1}{13}$
   $Y = 15$

8. $13(15) = 195$
   $195 = 195$

9. $\dfrac{25}{16} \div \dfrac{5}{8} = \dfrac{25}{\cancel{16}} \times \dfrac{\cancel{8}}{\cancel{5}} = \dfrac{5}{2} = 2\dfrac{1}{2}$

10. $\dfrac{19}{3} \div \dfrac{19}{8} = \dfrac{\cancel{19}}{3} \times \dfrac{8}{\cancel{19}} = \dfrac{8}{3} = 2\dfrac{2}{3}$

11. $5\dfrac{13}{8} - 3\dfrac{7}{8} = 2\dfrac{6}{8} = 2\dfrac{3}{4}$

12. $11\dfrac{5}{4} - 11\dfrac{3}{4} = \dfrac{2}{4} = \dfrac{1}{2}$

13. $1\dfrac{1}{3} + 1\dfrac{2}{3} = 2\dfrac{3}{3} = 3$

14. $2 \times 2{,}000 = 4{,}000$ lb

15. $24 \div 12 = 2$ ft

16. $\dfrac{1}{4}$ of $2{,}000 = 500$ lb

17. $\dfrac{3}{2} \times \dfrac{2000}{1} = \dfrac{6000}{2} = 3{,}000$ lb

18. $\dfrac{\cancel{5}}{\cancel{8}} \times \dfrac{1}{10} \times \dfrac{\cancel{4}}{\cancel{5}} = \dfrac{1}{20}$ of the money

19. $\dfrac{13}{2} \div \dfrac{13}{1} = \dfrac{\cancel{13}}{2} \times \dfrac{1}{\cancel{13}} = \dfrac{1}{2}$ dozen
    $\dfrac{1}{2}$ of $12 = 6$ doughnuts per person

20. $28 \div 4 = 7$
    $7 + 1 = 8$
    $8 \times 8 = 64$
    $64 - 4 = 60$

Lesson Test 26

1. $6B + 10 = 40$
   $6B = 30$
   $\dfrac{1}{6} \cdot 6B = 30 \cdot \dfrac{1}{6}$
   $B = 5$

2. $6(5) + 10 = 40$
   $30 + 10 = 40$
   $40 = 40$

3. $8N - 13 = 3$
   $8N = 16$
   $\dfrac{1}{8} \cdot 8N = 16 \cdot \dfrac{1}{8}$
   $N = 2$

4. $8(2) - 13 = 3$
   $16 - 13 = 3$
   $3 = 3$

5. $6C + 8 = 50$
   $6C = 42$
   $\dfrac{1}{6} \cdot 6C = 42 \cdot \dfrac{1}{6}$
   $C = 7$

6. $6(7) + 8 = 50$
   $42 + 8 = 50$
   $50 = 50$

7. $\dfrac{6}{\cancel{5}} \times \dfrac{\cancel{10}}{\cancel{3}} \times \dfrac{3}{11} = \dfrac{12}{11} = 1\dfrac{1}{11}$

8. $\dfrac{1}{\cancel{2}} \times \dfrac{5}{\cancel{12}} \times \dfrac{\cancel{24}}{7} = \dfrac{5}{7}$

9. $\dfrac{1}{5} \div \dfrac{8}{25} = \dfrac{1}{\cancel{5}} \times \dfrac{\cancel{25}}{8} = \dfrac{5}{8}$

10. $\dfrac{20}{3} \div \dfrac{20}{7} = \dfrac{\cancel{20}}{3} \times \dfrac{7}{\cancel{20}} = \dfrac{7}{3} = 2\dfrac{1}{3}$

11. $3 \times 5{,}280 = 15{,}840$ ft

12. $7 \times 16 = 112$ oz

13. $\dfrac{15}{4} \times 12 = \dfrac{180}{4} = 45$ in

14. $2\frac{1}{4} + 3\frac{1}{4} = 5\frac{2}{4} = 5\frac{1}{2}$ tons

15. no

16. $2 \times 2 \times 2 \times 2 \times 3$

17. 18 : $\underline{2}$, $\underline{3}$, $\underline{6}$, 9 , 18
    24 : $\underline{2}$, $\underline{3}$, 4 , $\underline{6}$, 8 , 12 , 24
    GCF = 6

18. $\frac{3}{2} \times \frac{3}{2} = \frac{9}{4} = 2\frac{1}{4}$ sq mi

19. $16 \times 16 = 256$

20. $56 \div 7 = 8$
    $8 + 10 = 18$
    $18 - 3 = 15$
    $15 + 15 = 30$

Lesson Test 27

1. $\frac{22}{7}\left(35^2\right) = \frac{22}{\cancel{7}} \times \frac{\cancel{1225}^{175}}{1}$
   $= \frac{3850}{1} = 3,850$ sq in

2. $\frac{2}{1} \times \frac{22}{\cancel{7}} \times \frac{\cancel{35}^{5}}{1} = \frac{220}{1} = 220$ in

3. $\frac{22}{7}\left(5^2\right) = \frac{22}{7} \times \frac{25}{1} = \frac{550}{7} = 78\frac{4}{7}$ sq ft

4. $\frac{2}{1} \times \frac{22}{7} \times \frac{5}{1} = \frac{220}{7} = 31\frac{3}{7}$ ft

5. $\frac{22}{7}\left(63^2\right) = \frac{22}{\cancel{7}} \times \frac{\cancel{3969}^{567}}{1}$
   $= \frac{12474}{1} = 12,474$ sq in

6. $\frac{2}{1} \times \frac{22}{\cancel{7}} \times \frac{\cancel{63}^{9}}{1} = \frac{396}{1} = 396$ in

7. $\frac{22}{7}\left(3^2\right) = \frac{22}{7} \times \frac{9}{1} = \frac{198}{7} = 28\frac{2}{7}$ sq ft

8. $\frac{2}{1} \times \frac{22}{7} \times \frac{3}{1} = \frac{132}{7} = 18\frac{6}{7}$ ft

9. $5R - 17 = 38$
   $5R = 55$
   $\frac{1}{5} \cdot 5R = 55 \cdot \frac{1}{5}$
   $R = 11$

10. $5(11) - 17 = 38$
    $55 - 17 = 38$
    $38 = 38$

11. $\frac{\cancel{4}^{2}}{3} \times \frac{3}{\cancel{2}} = \frac{2}{1} = 2$

12. $\frac{5}{\cancel{3}} \times \frac{7}{\cancel{3}} \times \frac{\cancel{9}^{3}}{\cancel{10}_{2}} = \frac{7}{2} = 3\frac{1}{2}$

13. $\frac{7}{10} \div \frac{7}{12} = \frac{\cancel{7}}{\cancel{10}_{5}} \times \frac{\cancel{12}^{6}}{\cancel{7}} = \frac{6}{5} = 1\frac{1}{5}$

14. $\frac{22}{5} \div \frac{7}{9} = \frac{22}{5} \times \frac{9}{7} = \frac{198}{35} = 5\frac{23}{35}$

15. 19 qt < 20 qt

16. 6,000 lb > 5,000 lb

17. 3,520 ft < 3,720 ft

18. $8\frac{4}{6} - 6\frac{3}{6} = 2\frac{1}{6}$ mi

19. $\frac{13}{6} \times 5,280 = \frac{68,640}{6} = 11,440$ ft

20. $\frac{1}{\cancel{2}} \times \frac{\cancel{4}^{2}}{1} \times \frac{5}{\cancel{2}} = \frac{5}{1} = 5$ sq in

Lesson Test 28

1. $\frac{4}{15}W + 7 = 19$
   $\frac{4}{15}W = 12$
   $\frac{15}{4} \cdot \frac{4}{15}W = \frac{12}{1} \cdot \frac{15}{4}$
   $W = \frac{180}{4}$
   $W = 45$

2. $\frac{4}{15}(45) + 7 = 19$
   $\frac{4}{15} \cdot \frac{45}{1} + 7 = 19$
   $\frac{180}{15} + 7 = 19$
   $12 + 7 = 19$
   $19 = 19$

3. $\frac{2}{3}R - 5 = 5$

$\frac{2}{3}R = 10$

$\frac{3}{2} \cdot \frac{2}{3}R = \frac{10}{1} \cdot \frac{3}{2}$

$R = \frac{30}{2} = 15$

4. $\frac{2}{3}(15) - 5 = 5$

$\frac{2}{3} \cdot \frac{15}{1} - 5 = 5$

$\frac{30}{3} - 5 = 5$

$10 - 5 = 5$

$5 = 5$

5. $\frac{2}{1} \cdot \frac{1}{2}X = \frac{4}{1} \cdot \frac{2}{1}$

$X = \frac{8}{1} = 8$

6. $\frac{1}{2} \cdot \frac{8}{1} = 4$

$\frac{8}{2} = 4$

$4 = 4$

7. $\frac{\overset{5}{\cancel{15}}}{\cancel{8}} \times \frac{\cancel{8}}{\cancel{3}} = \frac{5}{1} = 5$

8. $\frac{\cancel{4}}{5} \times \frac{\overset{5}{\cancel{10}}}{\cancel{3}} \times \frac{\overset{3}{\cancel{15}}}{\underset{2}{\cancel{8}}} = \frac{5}{1} = 5$

9. $\frac{11}{18} \div \frac{3}{8} = \frac{11}{\underset{9}{\cancel{18}}} \times \frac{\overset{4}{\cancel{8}}}{3} = \frac{44}{27} = 1\frac{17}{27}$

10. $\frac{9}{2} \div \frac{5}{2} = \frac{9 \div 5}{1} = \frac{9}{5} = 1\frac{4}{5}$

11. 48 oz > 32 oz

12. 1,056 ft = 1,056 ft

13. 1 qt < 2 qt

14. $2 + 4 + 6 + 8 = 20$

$20 \div 4 = 5$

15. $1 + 3 + 5 + 7 = 16$

$16 \div 4 = 4$

16. $8 + 10 + 12 = 30$

$30 \div 3 = 10$

17. $\frac{22}{7}(28^2) = \frac{22}{7} \cdot \frac{784}{1} = \frac{2464}{1} = 2{,}464$ sq yd

18. $\frac{2}{1} \times \frac{22}{\cancel{7}} \times \frac{\overset{4}{\cancel{28}}}{1} = \frac{176}{1} = 176$ yd

19. $\frac{11}{4} \times \frac{11}{4} \times \frac{11}{4} = \frac{1331}{64} =$

$20\frac{51}{64}$ cu in

20. $9 \times 8 = 72$

$72 - 2 = 70$

$70 \div 10 = 7$

$7 \times 4 = 28$

Lesson Test 29

1. $\dfrac{5}{5} = \dfrac{100}{100} = 100\%$

2. $\dfrac{3}{5} = \dfrac{60}{100} = 60\%$

3. $\dfrac{4}{5} = \dfrac{80}{100} = 0.80 = 80\%$

4. $\dfrac{3}{4} = \dfrac{75}{100} = 0.75 = 75\%$

5. $\dfrac{3}{5}Y + 2 = 26$

$\dfrac{3}{5}Y = 24$

$\dfrac{5}{3} \cdot \dfrac{3}{5}Y = \dfrac{24}{1} \cdot \dfrac{5}{3}$

$Y = \dfrac{120}{3} = 40$

6. $\dfrac{3}{5}(40) + 2 = 26$

$\dfrac{3}{5} \cdot \dfrac{40}{1} + 2 = 26$

$\dfrac{120}{5} + 2 = 26$

$24 + 2 = 26$

$26 = 26$

7. $6\dfrac{7}{5} - 2\dfrac{4}{5} = 4\dfrac{3}{5}$

8. $8\dfrac{4}{4} - 6\dfrac{3}{4} = 2\dfrac{1}{4}$

9. $10\dfrac{72}{80} + 4\dfrac{30}{80} = 14\dfrac{102}{80} = 15\dfrac{22}{80} = 15\dfrac{11}{40}$

10. 10

11. 14

12. 22

13. 35

14. III

15. XVII

16. XXV

17. XIX

18. $15 + 10 + 8 + 25 = \$58$

$\$58 \div 4 = 14\dfrac{2}{4} = \$14\dfrac{1}{2}$ per day

19. $A \approx \dfrac{22}{7}\left(14^2\right) = \dfrac{22}{\cancel{7}} \cdot \dfrac{\overset{28}{\cancel{196}}}{1}$

$= \dfrac{616}{1} = 616$ sq in

$C \approx \dfrac{2}{1} \cdot \dfrac{22}{\cancel{7}} \cdot \dfrac{\overset{2}{\cancel{14}}}{1} = \dfrac{88}{1} = 88$ in

20. $\dfrac{1}{2} \cdot \dfrac{16}{1} \cdot \dfrac{23}{4} = \dfrac{368}{8} = 46$ sq ft

Lesson Test 30

1. $\dfrac{4}{9}W + \dfrac{1}{3} = \dfrac{2}{3}$

$\dfrac{4}{9}W = \dfrac{1}{3}$

$\dfrac{9}{4} \cdot \dfrac{4}{9}W = \dfrac{1}{3} \cdot \dfrac{9}{4}$

$W = \dfrac{9}{12} = \dfrac{3}{4}$

2. $\dfrac{4}{9} \cdot \dfrac{3}{4} + \dfrac{1}{3} = \dfrac{2}{3}$

$\dfrac{12}{36} + \dfrac{1}{3} = \dfrac{2}{3}$

$\dfrac{1}{3} + \dfrac{1}{3} = \dfrac{2}{3}$

$\dfrac{2}{3} = \dfrac{2}{3}$

3. $\dfrac{1}{2}K - \dfrac{3}{4} = 4\dfrac{1}{4}$

$\dfrac{1}{2}K = 5$

$\dfrac{2}{1} \cdot \dfrac{1}{2}K = 5 \cdot \dfrac{2}{1}$

$K = 10$

4. $\dfrac{1}{2}(10) - \dfrac{3}{4} = 4\dfrac{1}{4}$

$5 - \dfrac{3}{4} = 4\dfrac{1}{4}$

$4\dfrac{1}{4} = 4\dfrac{1}{4}$

5. $\dfrac{1}{5} = \dfrac{20}{100} = 0.20 = 20\%$

6. $\dfrac{1}{4} = \dfrac{25}{100} = 0.25 = 25\%$

7. $\dfrac{15}{8} \div \dfrac{5}{8} = \dfrac{15 \div 5}{1} = 3$

8. $\dfrac{\overset{2}{\cancel{10}}}{8} \times \dfrac{\overset{2}{\cancel{16}}}{5} \times \dfrac{26}{5} = \dfrac{104}{5} = 20\dfrac{4}{5}$

9. $8\dfrac{12}{18} - 4\dfrac{15}{18} = 7\dfrac{30}{18} - 4\dfrac{15}{18} =$

$3\dfrac{15}{18} = 3\dfrac{5}{6}$

10. $11\dfrac{9}{18} + 6\dfrac{10}{18} = 17\dfrac{19}{18} = 18\dfrac{1}{18}$

11. 51

12. 115

13. 1,600

14. XI

15. CCC

16. MDCC

17. $5,280 \times 2 = 10,560$ ft

18. $2,000 \times 3 = 6,000$ lb

19. $\dfrac{15}{1} \times \dfrac{\overset{5}{\cancel{10}}}{1} \times \dfrac{7}{\cancel{2}} = 525$ cu ft

20. $8X - 8 = 40$

$8X = 48$

$\dfrac{1}{8} \cdot 8X = 48 \cdot \dfrac{1}{8}$

$X = 6$

Unit Test IV

1. $7L = 49$

$\dfrac{1}{7} \cdot 7L = 49 \cdot \dfrac{1}{7}$

$L = 7$

2. $7(7) = 49$

$49 = 49$

3. $3R - 8 = 22$

$3R = 30$

$\dfrac{1}{3} \cdot 3R = 30 \cdot \dfrac{1}{3}$

$R = 10$

4. $3(10) - 8 = 22$

$30 - 8 = 22$

$22 = 22$

5. $\dfrac{5}{8}X + 9 = 19$

$\dfrac{5}{8}X = 10$

$\dfrac{8}{5} \cdot \dfrac{5}{8}X = 10 \cdot \dfrac{8}{5}$

$X = \dfrac{80}{5} = 16$

6. $\dfrac{5}{8}(16) + 9 = 19$

$\dfrac{5}{8} \cdot \dfrac{16}{1} + 9 = 19$

$\dfrac{80}{8} + 9 = 19$

$10 + 9 = 19$

$19 = 19$

7. $\dfrac{3}{10}Y - \dfrac{1}{5} = \dfrac{2}{5}$

$\dfrac{3}{10}Y = \dfrac{3}{5}$

$\dfrac{10}{3} \cdot \dfrac{3}{10}Y = \dfrac{3}{5} \cdot \dfrac{10}{3}$

$Y = \dfrac{30}{15} = 2$

8. $\dfrac{3}{10}(2) - \dfrac{1}{5} = \dfrac{2}{5}$

$\dfrac{3}{10} \cdot \dfrac{2}{1} - \dfrac{1}{5} = \dfrac{2}{5}$

$\dfrac{6}{10} - \dfrac{1}{5} = \dfrac{2}{5}$

$\dfrac{3}{5} - \dfrac{1}{5} = \dfrac{2}{5}$

$\dfrac{2}{5} = \dfrac{2}{5}$

9. $\dfrac{3}{4}T + \dfrac{3}{8} = 6\dfrac{3}{8}$

$\dfrac{3}{4}T = 6$

$\dfrac{4}{3} \cdot \dfrac{3}{4}T = 6 \cdot \dfrac{4}{3}$

$T = \dfrac{24}{3} = 8$

10. $\frac{3}{4} \cdot \frac{8}{1} + \frac{3}{8} = 6\frac{3}{8}$

$\frac{24}{4} + \frac{3}{8} = 6\frac{3}{8}$

$6 + \frac{3}{8} = 6\frac{3}{8}$

$6\frac{3}{8} = 6\frac{3}{8}$

11. $\frac{\overset{3}{\cancel{9}}}{\underset{2}{\cancel{10}}} \times \frac{5}{7} \times \frac{\cancel{2}}{\cancel{3}} = \frac{3}{7}$

12. $\frac{19}{\cancel{9}} \times \frac{\cancel{9}}{\cancel{4}} \times \frac{\overset{3}{\cancel{12}}}{11} = \frac{57}{11} = 5\frac{2}{11}$

13. $\frac{3}{4} = \frac{75}{100} = 0.75 = 75\%$

14. $\frac{1}{2} = \frac{50}{100} = 0.50 = 50\%$

15. $\frac{22}{7}\left(49^2\right) = \frac{22}{7} \cdot \frac{2401}{1} =$

$\frac{7546}{1} = 7{,}546 \text{ sq in}$

16. $\frac{2}{1} \cdot \frac{22}{\cancel{7}} \cdot \frac{\overset{7}{\cancel{49}}}{1} = \frac{308}{1} = 308 \text{ in}$

17. 10

18. 1

19. 50

20. 1,000

21. 500

22. 5

23. 100

24. $5{,}280 \times 5 = 26{,}400 \text{ ft}$

25. $\frac{13}{\cancel{2}} \times \frac{\overset{1000}{\cancel{2000}}}{1} = 13{,}000 \text{ lb}$

26. $\frac{23}{\cancel{4}} \times \frac{\overset{3}{\cancel{12}}}{1} = 69 \text{ in}$

27. $169 \div 12 = 14\frac{1}{12} \text{ ft}$

28. $100 + 90 + 85 + 97 = 372$

$372 \div 4 = 93 \text{ average score}$

Final Test

1. $24 \div 2 = 12$

$12 \times 1 = 12$

2. $18 \div 3 = 6$

$6 \times 2 = 12$

3. $64 \div 8 = 8$

$8 \times 7 = 56$

4. $\frac{3}{4} = \frac{6}{8} = \frac{9}{12} = \frac{12}{16}$

5. $\frac{9}{10} = \frac{18}{20} = \frac{27}{30} = \frac{36}{40}$

6. $\frac{25}{35} > \frac{21}{35}$

7. $\frac{24}{48} = \frac{24}{48}$

8. $\frac{12}{24} < \frac{16}{24}$

9. $\frac{3}{9} + \frac{5}{9} = \frac{8}{9}$

10. $\frac{4}{8} + \frac{2}{8} + \frac{7}{8} = \frac{13}{8} = 1\frac{5}{8}$

11. $\frac{12}{15} - \frac{5}{15} = \frac{7}{15}$

12. $\frac{1}{3} \times \frac{5}{1} = \frac{5}{3} = 1\frac{2}{3}$

13. $\frac{\overset{2}{\cancel{10}}}{\cancel{3}} \times \frac{\overset{6}{\cancel{18}}}{\cancel{5}} = 12$

14. $\frac{\cancel{19}}{5} \times \frac{\overset{5}{\cancel{25}}}{\underset{3}{\cancel{57}}} = \frac{5}{3} = 1\frac{2}{3}$

15. $7\frac{1}{4} - 5\frac{3}{4} = 6\frac{5}{4} - 5\frac{3}{4} = 1\frac{2}{4} = 1\frac{1}{2}$

16. $9\frac{2}{3} + 6\frac{5}{9} = 9\frac{18}{27} + 6\frac{15}{27}$

$= 15\frac{33}{27} = 16\frac{6}{27} = 16\frac{2}{9}$

17. $5\frac{1}{5} - 2\frac{5}{6} = 5\frac{6}{30} - 2\frac{25}{30}$

$= 4\frac{36}{30} - 2\frac{25}{30} = 2\frac{11}{30}$

18. $2\frac{5}{8} \text{ in}$

19. $7X + 9 = 44$

$7X = 35$

$\frac{1}{7} \cdot 7X = 35 \cdot \frac{1}{7}$

$X = 5$

20. $7(5) + 9 = 44$

$35 + 9 = 44$

$44 = 44$

21. $\frac{3}{8}A - 8 = 13$

$\frac{3}{8}A = 21$

$\frac{8}{3} \cdot \frac{3}{8}A = 21 \cdot \frac{8}{3}$

$A = \frac{168}{3} = 56$

22. $\frac{3}{8}(56) - 8 = 13$

$21 - 8 = 13$

$13 = 13$

23. $\frac{5}{6}G + \frac{1}{6} = \frac{5}{12}$

$\frac{5}{6}G = \frac{5}{12} - \frac{1}{6}$

$\frac{5}{6}G = \frac{5}{12} - \frac{2}{12}$

$\frac{5}{6}G = \frac{3}{12} \text{ or } \frac{1}{4}$

$\frac{6}{5} \cdot \frac{5}{6}G = \frac{1}{4} \cdot \frac{6}{5}$

$G = \frac{6}{20} = \frac{3}{10}$

24. $\frac{5}{6} \cdot \frac{3}{10} + \frac{1}{6} = \frac{5}{12}$

$\frac{15}{60} + \frac{1}{6} = \frac{5}{12}$

$\frac{1}{4} + \frac{1}{6} = \frac{5}{12}$

$\frac{5}{12} = \frac{5}{12}$

25. $\frac{\cancel{5}}{8} \times \frac{1}{\cancel{3}} \times \frac{\cancel{3}}{\cancel{5}} = \frac{1}{8}$

26. $\frac{\cancel{4}}{5} \times \frac{11}{\cancel{4}} \times \frac{\overset{2}{\cancel{10}}}{3} = \frac{22}{3} = 7\frac{1}{3}$

27. $\frac{4}{5} = \frac{80}{100} = 0.80 = 80\%$

28. $\frac{1}{4} = \frac{25}{100} = 0.25 = 25\%$

29. $15: \underline{3}, \underline{5}, \underline{15}$

$45: \underline{3}, \underline{5}, 9, \underline{15}, 45$

$GCF = 15$

30. $2 \times 2 \times 2 \times 7$

31. $7\frac{2}{3} = \frac{23}{3}$

32. no

33. $\frac{22}{7}(21^2) = \frac{22}{\cancel{7}} \cdot \frac{\overset{63}{\cancel{441}}}{1} = \frac{1386}{1} = 1{,}386 \text{ sq ft}$

34. $\frac{2}{1} \cdot \frac{22}{\cancel{7}} \cdot \frac{\overset{3}{\cancel{21}}}{1} = 132 \text{ ft}$

Multistep Problems Lesson 6

1. area of garden: $25 \times 15 = 375$ sq ft

   $\frac{1}{3} \times 375 = 125$ sq ft for corn

   $\frac{1}{5} \times 375 = 75$ sq ft for peas

   $125 + 75 = 200$ sq ft used

   $375 - 200 = 175$ sq ft left over

2. $\frac{1}{2} \times 32 = 16$ signed by Sarah

   $\frac{3}{8} \times 32 = 12$ signed by Richard

   $16 + 12 = 28$ total signed

   $32 - 28 = 4$ cards left to be signed

3. $\$15 + \$9 + \$12 = \$36$ needed

   $\frac{1}{4} \times \$36 = \$9$ on hand

   $\$36 - \$9 = \$27$ to earn

Multistep Problems Lesson 12

1. $\frac{1}{5} \times \$30 = \$6$ for sandwich

   $\frac{1}{6} \times \$30 = \$5$ for museum

   $\frac{1}{2} \times \$30 = \$15$ for book

   $\$6 + \$5 + \$15 = \$26$ spent

   $\$30 - \$26 = \$4$ left over

2. $\frac{1}{2} \times 128 = 64$ miles Mark drove

   $\frac{1}{4} \times 128 = 32$ miles Justin drove

   $64 + 32 = 96$ miles driven

   $128 - 96 = 32$ miles left

   $32 \div 2 = 16$ miles driven by Kaitlyn

3. $\frac{1}{2} + \frac{3}{4} = \frac{2}{4} + \frac{3}{4} = \frac{5}{4}$ pie

   $\frac{5}{4} \div \frac{1}{8} = \frac{40}{32} \div \frac{4}{32} =$

   $40 \div 4 = 10$ pieces of pie

Multistep Problems Lesson 18

1. $63 + 21 + 63 + 21 = 168$ yd ▭▭▭
   for outside of field

   $21 + 21 = 42$ yd for dividing sections

   $168 + 42 = 210$ yd of fencing

   $210 \times 3 = 630$ ft of fencing

2. $\frac{1}{2} \times 1{,}824 = 912$ with team colors

   $\frac{2}{3} \times 912 = 608$ with hats and colors

   $\frac{1}{4} \times 608 = 152$ with hats, colors, and banners

   $152 + 100 + 25 = 277$ with banners

3. $36" \div 3 = 3$ ft

   $2 \times 3 = 6$ ft to brim, so cube is 6 ft on a side

   $6 \times 6 \times 6 = 216$ cu ft volume

   $216 \times 63 = 13{,}608$ pounds of water

Multistep Problems Lesson 24

1. $2\frac{1}{2} + 1\frac{1}{8} + 1\frac{1}{4} = 4\frac{7}{8}$ pounds of salad

   $4\frac{7}{8} - 1\frac{1}{2} = 3\frac{3}{8}$ pounds left

   $3\frac{3}{8} \div 3 = \frac{27}{8} \div \frac{3}{1} = \frac{27}{8} \times \frac{1}{3} =$

   $\frac{9}{8} = 1\frac{1}{8}$ lb in a container

2. $1 \times 1 \times 2 = 2$ cu ft in aquarium

   $3 \times \frac{1}{4} = \frac{3}{4}$ cu ft for one kind of fish

   $2 \times \frac{1}{3} = \frac{2}{3}$ cu ft for other kind of fish

   $\frac{3}{4} + \frac{2}{3} = 1\frac{5}{12}$ cu ft for both kinds

   $2 - 1\frac{5}{12} = \frac{7}{12}$ cu ft

   $\frac{2}{3} = \frac{8}{12}$, so there is not room for extra fish.

3. $5\frac{1}{2} \div 1\frac{1}{2} = \frac{11}{2} \times \frac{2}{3} = \frac{22}{6} = 3\frac{4}{6} = 3\frac{2}{3}$ pieces

   $1\frac{1}{2}$ yd $\times 36$ in $= 54$ in: length of one piece

   $\frac{2}{3} \times 54 = 36$ in: length of leftover piece

Symbols and Tables

SYMBOLS

=	equals
≈	approximately equals
<	less than
>	greater than
%	percent
π	pi (approximately 22/7 or 3.14)
r^2	r squared, or r ·r
'	foot
"	inch
⌐	right angle

PERIMETER

For any figure with straight sides, add the lengths of all the sides.

CIRCUMFERENCE

For a circle: C = 2πr

AREA

For a rectangle, square, or parallelogram:
A = bh (base times height)

For a triangle: $A = \dfrac{bh}{2}$

For a circle: $A = \pi r^2$

For a trapezoid: $A = \dfrac{b_1 + b_2}{2} \times h$

(average of two bases times height)

EXPANDED NOTATION

$1,452.5 = 1 \times 1,000 + 4 \times 100 + 5 \times 50 + 2 \times 1 + 5 \times 1/10$

MEASUREMENT

3 teaspoons (tsp) = 1 tablespoon (Tbsp)

2 pints (pt) = 1 quart (qt)

8 pints (pt) = 1 gallon (gal)

4 quarts (qt) = 1 gallon (gal)

12 inches (in) = 1 foot (ft)

3 feet (ft) = 1 yard (yd)

5,280 feet (ft) = 1 mile (mi)

16 ounces (oz) = 1 pound (lb)

2,000 pounds (lb) = 1 ton

60 seconds = 1 minute

60 minutes = 1 hour

7 days = 1 week

365 days = 1 year (366 days in leap year)

52 weeks = 1 year

12 months = 1 year

100 years = 1 century

1 dozen = 12

DIVISIBILITY RULES

A number is divisible by:

2	if it ends in an even number.
3	if the digits add to a multiple of 3.
5	if it ends in 5 or 0.
9	if the digits add to a multiple of 9.
10	if it ends in 0.

VOLUME

For a rectangular solid: V = Bh
(area of base times height)

For a cube: V = Bh

Glossary

A–D

Arabic numerals - the numerals commonly used today

Area - the amount of space a two-dimensional shape occupies

Average - the result of adding a series of numbers and dividing by the number of items in the series

Base - the top or bottom side of a shape

Canceling - dividing a numerator and a denominator by the same number in order to simplify the problem

Century - one hundred years

Circumference - the distance around a circle; corresponds to perimeter

Coefficient - in this book, a number directly in front of an unknown. It is multiplied by the unknown.

Composite number - a number with factors other than one and itself

Cube - a three-dimensional figure with each edge the same length

Cubic units - units multiplied by the same units three times

Decimal or decimal fraction - a fraction written on one line by using a decimal point and place value

Decomposition - the process of breaking a number down into parts that can be re-combined to find the total

Denominator - the bottom number in a fraction. It tells how many total parts are in the whole.

Dimension - a measurement in a particular direction used to find area or volume

Dividend - the number that is being divided in a division operation; represents the total

Divisor - the number that the dividend is being divided by in a division operation

E–G

Equation - a number sentence where the value of one side is equal to the value of the other side

Equivalent - having the same value

Estimation - a skill used to get an approximate value of an answer

Even number - an integer which is a multiple of two. When written in base ten, even numbers end with the digits 0, 2, 4, 6, or 8.

Expanded notation - a way of writing numbers where each digit is multiplied by its place value

Exponent - a small raised number that indicates how many times another number is to be used as a factor

Factors - the numbers being multiplied in a multiplication operation

Factor tree - a method for finding the prime factors of a given number

Fraction - a number described using an expression that consists of a numerator divided by a denominator

Greatest common factor (GCF) - the greatest number that is a factor of two or more given numbers

H–P

Height - the length of a line segment perpendicular to the base of a shape which passes through the point farthest from the base

Improper fraction - a fraction with a numerator greater than its denominator

Inequality - a number sentence where the value of one side is greater than or less than the value of the other side

Mixed number - a number made up of a whole number and a fraction

Multiplicative inverse - a number that, when multiplied times the original number, equals one; also referred to as the reciprocal

Numerator - the top number in a fraction. It tells how many of the parts of a whole are being considered.

Odd number - an integer which is not a multiple of two. When written in base ten, odd numbers end with the digits 1, 3, 5, 7, or 9.

Percent - "per hundredth" - a way of writing fractions that have a denominator of 100. The symbol is %.

Pi - the number that results when we divide the circumference of any circle by the radius of that circle. It is represented by the Greek letter π and is slightly less than 3 1/7 or 22/7. In *Epsilon* we will use the value 22/7 to find the approximate area and circumference of a circle.

Place value - the position of a digit that tells its value

Prime factor - the factor of a number that can only be divided by one and itself

Prime number - a whole number greater than one with only two factors, one and itself

Product - the result of a multiplication operation

Proper fraction - a fraction with a numerator less than its denominator

Quadrilateral - a closed two-dimensional shape with exactly four straight sides

Quotient - the result of a division operation

R–S

Radius - the distance from the center of a circle to its edge

Reciprocal - a number that, when multiplied by the original number, equals one. The reciprocal of a fraction is found by switching the positions of the numerator and denominator.

Rectangle - a quadrilateral which has four "square corners," or right angles

Rectangular solid or rectangular prism - a three-dimensional shape with six rectangular faces

Regrouping - converting numbers from one place value to another in order to solve a problem

Repeated division - a method for finding the prime factors of a given number

Right angle - a square corner (90° angle)

Roman numerals - a numbering system used by the Roman Empire and many other people. Instead of place value, it requires adding and subtracting symbols that represent different numbers to get a total.

Rounding - describing a number using a nearby multiple of ten. Sometimes numbers are rounded in order to estimate the result of an operation.

"Rule of four " - a Math-U-See method for finding the common denominator of two fractions

"Same difference theorem" - the Math-U-See idea that we can add the same number to both the subtrahend and minuend in a subtraction expression without changing the difference

Simplifying - finding an equivalent fraction that has a numerator and denominator that cannot be divided by any common factor other than one. We can simplify a fraction by dividing the numerator and denominator of that fraction by the same number.

Square - a rectangle with four straight sides that have the same length

Square units - units multiplied by themselves. These can be thought of as equivalent to the area of a square.

T–Z

Trapezoid - a quadrilateral that has exactly one pair of parallel sides

Triangle - a closed shape that has exactly three straight sides

Unit fraction - a fraction with a numerator of one

Units - the first place value in the decimal system; also, a quantity used to measure

Unknown - a number whose value we do not know. It is usually represented by a letter of the alphabet.

Volume - the amount of space a three-dimensional shape occupies

Master Index for General Math

This index lists the levels at which main topics are presented in the instruction manuals for *Primer* through *Zeta*. For more detail, see the description of each level at MathUSee.com. (Many of these topics are also reviewed in subsequent student books.)

Epsilon Index